PROTEOMICS IN FOOD SCIENCE

PROTEOMICS IN FOOD SCIENCE

Abeer Iqbal

Delve
PUBLISHING

www.delvepublishing.com

Proteomics in Food Science

Abeer Iqbal

Delve Publishing

2010 Winston Park Drive,

2nd Floor

Oakville, ON L6H 5R7

Canada

www.delvepublishing.com

Tel: 001-289-291-7705
001-905-616-2116

Fax: 001-289-291-7601

Email: orders@arclereducation.com

ISBN: 978-1-77407-291-2 (Hardcover)

ABOUT THE AUTHOR

Abeer Iqbal is pursuing her PhD in Biological Sciences from University of The Punjab in Pakistan. She is also MPhil in Microbiology and Molecular Genetics from University of The Punjab, Pakistan. She is passionate about writing and research related to Biological Sciences. She has been working as an academic writer and medical blogger for the past few years and enjoy every part of her job, She feels that Genetics and Biochemistry are the real essence of life and learning and unraveling new researches in these areas is really amazing !!

TABLE OF CONTENTS

LIST OF FIGURES

Figure 10.4: The process of Mass Spectrometry has been one of the most frequently used methods in the field of Proteomics

PREFACE

Food is the most essential part of life as it provides the energy and nutrients that are necessary to survive and run the body. All the creatures depend on food to exist, especially humans, as it controls the way they grow and develop. Food is extremely essential for the body to function and carry out daily activities.

To control the way food behaves and to know about the various nutrients it provides to the humans, the field of food science was developed. Food science reveals the basic information on the contents of the food products and the value of the energy or the various levels of nutrition the product provides. The study of food science is helpful in recognizing the possibilities that may arise out of it and can change the scenario of food nutrition in the world.

The protein component of the food products forms a significant part of them as it is used in building the muscles in a body. The proteins help to increase the activity of the organisms at the cellular level that enables them to provide the energy to the bodywork on building the strength it requires.

Proteomics is the branch of science that deals with the study of proteins at the cellular level. Proteomics provides the answers that the researchers are looking for, to improve the quality of food that is being consumed by humans.

This book informs the readers about the branch of proteomics, elaborates upon the importance it holds for the humans and their nutrition and shows how it can be helpful in the production and improvement of food and the related products.

The book starts off by introducing the subject of proteomics to the readers. It elaborates on its origin and the way proteomics came into existence. The book talks about the importance of the study of proteomics and how it can be utilized in various fields.

Proteomics is a scientific procedure that involves a lot of techniques and methods to be conducted in order to have a good understanding of the subject being concentrated on. The book explains the readers about the procedures that are undertaken in proteomics and the objective they serve. It also discusses the modern and upcoming techniques in the field of proteomics.

The book goes on to explain the meaning of plant proteomics to the readers. It discusses the potential of proteomics in the plant world as a tool of biotechnology. The book throws light on the recent technologies that have been innovative in their approach towards plant improvement. The book talks about the characterization of plant complexes and compounds through mass spectrometry. It also talks about the approaches and challenges in the field of plant proteomics.

The book also brings the focus of the readers to the use of proteomics in aquaculture. It emphasizes on the need of involving proteomics to improve the quality of the products from aquaculture. It stresses the importance of proteomics to assess the impacts of climate change on the aquatic organisms and how it changes the quality of products. It also explains to the readers how proteomics can be used to assess the potential risks associated with aquaculture.

The readers are informed about the various applications and uses of proteomics in food science. The book talks about the use of proteomics for the verification and validation of the processes in aquaculture. It discusses identifying the pathogens that may be borne due to food and the allergens that maybe there in the food products. It discusses the application of proteomics in the meat industry and also brings to notice of the readersthe future applications of proteomics.

1

Introduction to Proteomics

CONTENTS

Proteomics is known to be a branch of biology that deals with the study of the structural and functional analysis of proteins. The information that is obtained from a genome of an organism is known to be incomplete for understanding the molecular mechanisms in living organism. Proteins are known to play an important role in the cells of the living organisms because of which the study of proteins is extremely important to get a comprehensive idea regarding the living organism. In this book, different concepts related to the proteomics have been explained, several latest technologies that are used in the study of the proteins and these technologies have been mentioned. The different stages that are involved in the study of the proteins and the advantages and complexities that are present in the area of proteomics have been explained. We also attempt to describe various applications of the proteomics in drug discovery and the role and importance of proteomics has been explained along with the role of proteomics in food science.

1.1 INTRODUCTION TO PROTEOMICS

Proteomics is known to be a branch of molecular biology that is concerned with the study of the proteome of the organism. Proteomics is an emerging area in science which was initially introduced as a branch in the year 1994.

Proteins are known to be molecules that are completely associated with disease because of which the study of proteins is known to provide a lot of information regarding the diseased cell or tissue. The researchers indicate that the study of the proteome is extremely dynamic because of which the proteome is generally called as the "functional proteome".

The term functional proteome is known to refer to of several proteins that are produced in a cell within a stipulated time frame.

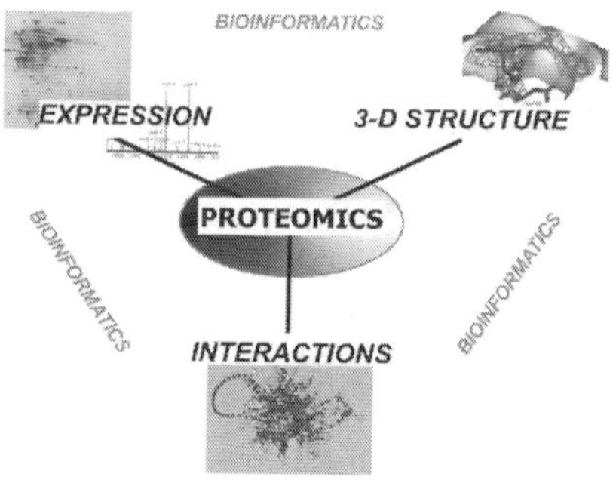

Figure 1.1: Illustration of Proteomics.

Source: https://upload.wikimedia.org/wikipedia/commons/1/14/Proteomics.jpg

Whilst humans are estimated to have 30,000 to 40,000 genes potentially encoding 40,000 different proteins, alternative RNA splicing and post-translational modification (PTM) may increase this number up to 2,000,000 proteins or protein fragments. This indicates that the proteome is way complicated than the genome of the cell.

Genomics is known to be an important field of science which deals with the entire genome of living organisms. The branch of genomics is known to be extremely popular in providing information related to the functions of the genes in an organism, but the genomic studies fail to indicate the changes that happen in the protein during post-translational modifications.

Researches indicate that it is extremely easy to work with the nucleic acids such as DNA/RNA whereas there are several limitations with respect to the study of information from the nucleic acids because of which study of proteins is known to be extremely popular.

There are several reasons for the incompleteness of the information that is obtained from the gene sequence. The lack of adequate information regarding the accuracy of the genomic profile leads to an increased importance for the proteomic profile of an organism. The protein molecules are known to provide a lot of information regarding the structure and function of a living organism.

After synthesis on ribosomes, proteins are cut to eliminate initiation, transit, and signal sequences, as well as simple chemical groups or complex molecules that are attached. The transcript of a gene is known to be spliced in several different ways before the translation of a gene into protein. Most of the proteins are generally chemically altered because of several post-translational modifications by the addition of different carbohydrate and phosphate groups.

This chemical change is known to play an important role in modulating the function of several proteins but these are not directly coded by the genes. There are several different types of post-translational modifications that occur in a cell and these post-translational modifications are known to static or dynamic. There are nearly 300 different types of post-translational modifications that have been identified.

The different types of post-translational modifications are glycosylation, phosphorylation, acetylation, deamidation, palmitoylation, sulfation and several others. The presence of different types of post-translational modifications is a reason for the occurrence of nearly 50 different protein sequences from a single gene.

This is the reason because of which the analysis of nucleic acids is known to be insufficient for the prediction of the different forms of proteins. There are several proteins that are formed because of the post-translational modification and mRNA splicing of the protein. The RNA quantization is inefficient in indicating the levels of the protein in a cell. There are several proteins that must be considered while PTM and mRNA splicing of a cell are considered.

DNA/RNA analysis cannot predict the amount of a gene product, including the type and amount of PTM when a gene is translated, or the events involving multiple genes such as aging, stress responses, drug responses, and pathological transformations. This indicates the reason because of which the genomic information of a protein is not enough for the appropriate information regarding the protein abundance, activity and structure of the protein.

The genomic analysis is not capable of providing exact details regarding several aspects of the living organisms because of which several proteins are being used for appropriate analysis of proteins. The analysis of proteins is known to provide a lot of information regarding the availability of the protein, structure of the proteins and activity of the proteins.

Genomics and proteomics are known to be complementary fields, but the field of proteomics can provide information regarding the functional analysis of several proteins. The researchers and scientists believe that the genomic and proteomic studies of a living organism can be instrumental in designing novel disease markers and drug targets for several living organisms.

These studies would play an important role in prevention, diagnosis and treatment for the disease. The future of different areas such as medicine and biotechnology are known to be greatly impacted because of genomics and proteomics.

Proteome refers to all the proteins that are coded by the genes of the organism. Proteomics generally refers to the entire protein content that is present in the cell of an organism. It plays an important role in understanding several changes that occur during the process of gene expression in the entire life span of an organism. There would be changes in the pattern of the gene expression of an organism because of changes in environmental conditions or changes in the stress of an organism.

Proteomics is known to play an important role in understanding the structure and function of several proteins. The study of proteomics is also known to play an important role in understanding the protein-protein interactions of an organism. It plays an important role in understanding several small defects in the structure of the proteins along with understanding the function or alternation in the expression pattern of several proteins.

The expression pattern of any genome is easily identified in several proteomic studies. The understanding of the expression pattern of any genome is extremely important in several drug development studies and structural analysis of proteins. Proteomic studies are vital in drug development as protein molecules act as promising targets for a wide range of diseases.

1.2 IMPORTANT CONCEPTS THAT ARE RELATED TO PROTEOMICS

There are some common terminologies that are commonly used in proteomics and some of these terms are mentioned below and are explained in detail. They are:

- Genome
- Genomics
- Replication
- Transcription
- Translation
- Transcriptome
- Transcriptomics
- Polymorphism
- Proteome
- Proteomics
- Metabolome
- Metabonomics

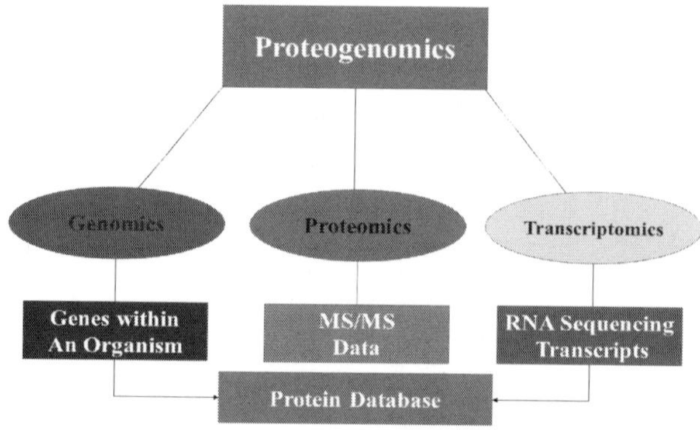

Figure 1.2: Different Fields Related to Proteomics.

Source: https://upload.wikimedia.org/wikipedia/commons/3/3a/Proteogenomics_Image_2.png

1.2.1 Genome

Genome refers to the entire quantity of the deoxyribose nucleic acid (DNA) that is present within the cell or a living organism.

1.2.2 Genomics

Genomics refers to the study of the structure, function and expression of several genes that are present in the living organism.

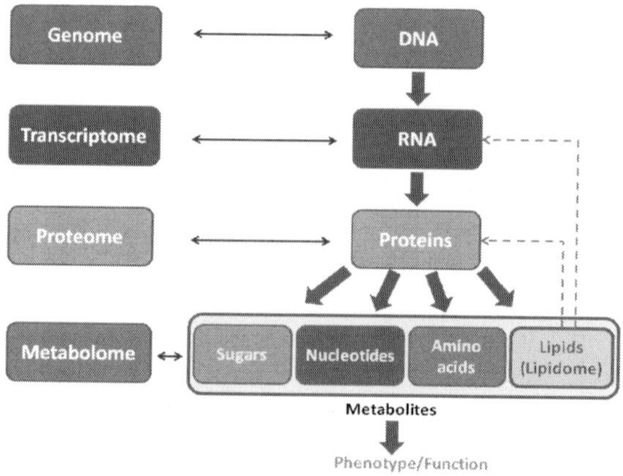

Figure 1.3: Concepts Related to Proteomics.

Source: https://upload.wikimedia.org/wikipedia/en/9/98/Metabolomics_schema.png

1.2.3 Replication

. The synthesis of an exact copy of the nucleotides from the DNA molecules is called as Replication.

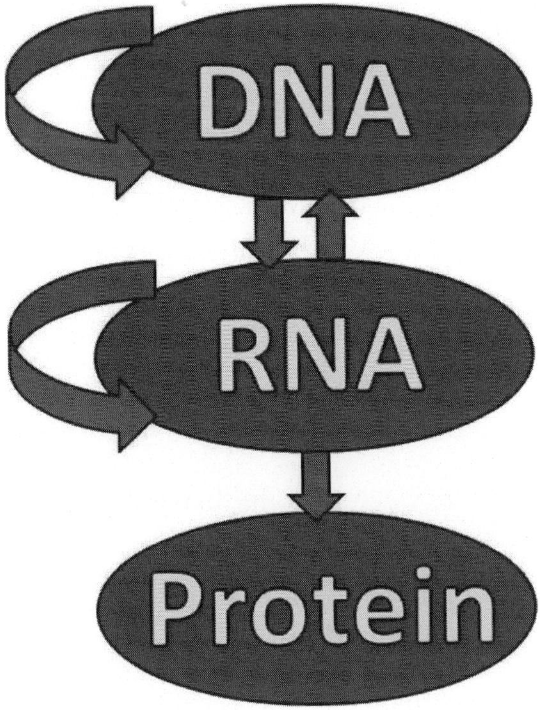

Figure 1.4: Process of Expressing Proteins.

Source: https://upload.wikimedia.org/wikipedia/commons/thumb/9/9e/Central_dogma.JPG/483px-Central_dogma.JPG

1.2.4 Transcription

The synthesis of several novel proteins generally begins with the process of transcription. In the process of transcription, the DNA molecule is transcribed to three types of RNA molecules, mRNA molecule called as the messenger RNA, rRNA which is called as ribosomal RNA or the tRNA which is the transport RNA.

This happens in the nucleus by means of enzymatic complexes produced by specific genes. This RNA that is produced is taken outside the nucleus for translation into proteins.

1.2.5 Translation

The translation is the process that involves the synthesis of proteins. The translation of a molecule generally happens in the cytoplasm of the cell. This process is known to convert the mRNA molecules into proteins based on the coding information that is present in the genes.

1.2.6 Transcriptase

The entire quantity of the mRNA that is present within the cell of a living organism is generally called Transcriptase.

1.2.7 Transcriptomics

Transcriptomics refers to the branch of science which deals with the study and research on the mRNA molecules that are present within a cell or a living organism.

1.2.8 Polymorphism

Polymorphism refers to the changes that occur at a specific site in DNA.

1.2.9 Proteome

The entire pool of proteins that are expressed by the genome of an organism that is present within the cell, tissue or organism is called proteome.

1.2.10 Proteomics

The large-scale study of the proteins that are present within a living organism along with the study of the structure, function, and analysis of the proteins is called proteomics. The term proteomics was called as an analogy to the term genomics.

1.2.11 Metabolome

The total quantitative collection of different types of several low molecular compounds known as metabolites that are present within a cell of an organism that participates in several metabolic reactions.

The metabolome includes the different metabolites that are absorbed from the external environment into a living organism and from the symbiotic relationships that are present in the living organism.

1.2.12 Metabonomics

A measure of the fingerprint of biochemical perturbations caused by disease, drugs, and toxins; some would say that metabonomics and metabolomics are the same and the terms are occasionally used interchangeably.

1.3 SHIFT FROM GENOMICS TO PROTEOMICS

The completion of the human genome project is shifting the focus from the genomic perspective of the organism to the protein complement of any living organism. This shift in attention has led to the arrival of a new area of science called proteomics. Proteomics is an emerging area in science that deals with the study of several novel aspects of a living organism related to the proteins.

The study of different proteins that are present in the cells of the living organism along with the functional study and quantification of proteins is called proteomics. Proteomics deals with the area which will deal with identification and detection of proteins along with the determination and understanding of the structures, functions and pathophysiological functions of several proteins.

The recent studies and research led to the completion of the analysis of the sequencing of the entire genome of several prokaryotic and eukaryotic organisms have increased the interest of the scientists and researchers in the area of proteomics.

The term proteome refers to the different types of proteins that are mostly expressed by the genome and the genome is known to be present undisturbed while it produces several different types of proteins. The proteins that are expressed and coded by the genome are known to be important in several physiological and biochemical functions of a living organism.

The genes produce different types of proteins based on the environment as they are known to have an internal turn on or turn off mechanism that is dictated by the surroundings in which that organism is present. Protein is known to be directly involved in different aspects of the functioning of the organisms.

1.3.1 Role and Scope of Proteomics

Proteomics is now an important branch of science which is considered with several different aspects of the protein and the major focus of this area is as follows:

- Proteomics is concerned with understanding and analyzing the structure and function of the protein
- Proteomics is a branch that is concerned with the study of the 3D structure of the protein and the study of the 3D structure of the protein plays an important role in drug designing.
- Proteomics is a branch that is concerned with the study of qualitative and quantitative analysis of proteins

1.4 DIFFERENT BRANCHES OF PROTEOMICS

Proteomics is a very vast branch of science and it is categorized into three types. The different types of proteomics are as follows:

- Functional Proteomics
- Structural Proteomics
- Differential Proteomics

Proteomics is organized into three different branches and they are as follows:

- Structural Proteomics
- Expression Proteomics
- Interaction proteomics

1.4.1 Structural Proteomics

Structural proteomics is a branch of proteomics that is concerned with the study of the interaction of several genes and drug interaction.

1.4.2 Expression Proteomics

Proteomics deals with the study of the main gene in a sample of proteins.

1.4.3 Interaction Proteomics

This area is the study of proteomics that deals with the study of the metabolic pathways by which proteins are converted into large complexes of proteins.

1.5 ADVANTAGES OF PROTEOMIC STUDIES

- Proteomics is a field of science that deals with the study of different proteins which give an insight that the genes are not the only cause for any disease and proteins expressed by these genes play an important role.
- Proteomics is a branch of science that helps in diagnosing the right treatment for diseases.
- The studies and research in several areas concerned to proteomics revealed that the proteomics plays an important role in understanding that the enzyme HIV protease is majorly responsible for causing the disease AIDS.
- An important application of proteomics is it is important in diagnosing of several diseases with the help of several biomarkers. There is another branch of proteomics that plays an important role in the study of the drugs in proteomics and it is called as pharmaco-proteomics.

1.6 COMPLEXITIES IN PROTEOMIC STUDIES

There are different types of complexities and problems that are associated with the study of proteins while comparing it to the study of nucleic acids. Some of the major complexities and problems that are involved with the working of proteins are given below:

- Proteins are extremely difficult to work with when compared to working with nucleic acids such as DNA and RNA. The major aspect related to the complexity of the protein is because of the secondary and tertiary structure of the proteins needs to be retained for the analysis of the proteins. Proteins are known to be denatured because of exposure to heat, light, enzymes and aggressive beating. These reasons aid to the study and research of several aspects related to
- The study of proteomics is an extremely difficult field because the concentration of the proteins is extremely variable, and it differs from one organism to the other and one cell in an organism to the other cell in the same organisms. This variable concentration of

proteins in different organisms makes it extremely complex to study proteomics.

- Proteins are compounds that have very poor solubility. The presence of extremely poor solubility makes it difficult to analyze and study proteins.
- The other aspect pertaining to the complexities of proteomics is that the proteins cannot be amplified easily like DNA molecules because of which it becomes extremely difficult to detect the proteins.
- Most of the protein content in the plasma of the cells generally comes from the protein albumin and from some other 10 proteins which constitute to nearly 90% of the total content of the cell.
- There are several proteomics technologies that are available to analyze different types of proteins and most of these technologies are used to analyze protein to nearly 3-4 orders of magnitude and these are generally analyzed at a higher concentration at one end of the spectrum.

The important step before analyzing proteins is to analyze several high abundant proteins from the plasma or serum as the removal of these proteins enables in conducting detailed analysis of proteins. This also plays an important role in reducing the complexities that are involved with the analysis of proteins.

However, the removal of highly abundant proteins like albumin is known to have several other losses because there would be loss of biomarkers that are attached to the highly abundant proteins. This would lead to a lot of other problems that are associated with working of proteins.

- There has been a novel method that has been designed to work with the low abundant proteins. This method would use a deep proteome approach through the ligand library beads. This method was initially used by *Thulasiraman et al.* The use of different types of equalizer beads along with the combinatorial library of ligands is an important method that is used to work with several low abundant proteins.

This process of discovery of novel methods is providing an approach to work with several proteins that would help in working with unseen proteome and the process of developing novel technologies to deal with a constant challenge that is arising is a process that deals with the process of working with these proteins.

- The major difficulty of working with the proteins is that the proteins are extremely sensitive as they are thermolabile and sensitive. Proteins can denature easily this problem is solved by the development of several ultra-microarrays that are developed.

These ultra-microarrays are known to have advantages of the microarrays that include the multiplexing capabilities, high throughput technologies and are economics to screen small amount of samples. These proteomic technologies are known to be useful in screening several samples that are obtained by microdissection, forensic samples, and neonatal biopsy micro specimens.

- There are several defined and known problems that are present in dealing with several aspects of proteomics. These technical obstacles are known to be important as they may be useful in the standardization of these methodologies and dissemination of the proteomics data.
- The major constraint with respect to the development of these proteomics technologies is the cost factor. These methods are extremely expensive, and the cost factor is an important constraint in the development of several new methods and techniques.
- There is another important challenge with respect to proteomics is the problems that are present because of complex instrumentation, critical computing power and expensive consumables that are present in the integration of technologies of proteomics and metabonomics. The standardization of several procedures and these technologies is an important challenge for people working in the area of proteomics.
- Another major challenge will be the integration of proteomics with genomics and metabolomics data, as well as their functional interpretation in conjunction with clinical results and epidemiology.

1.7 PROCEDURE USED IN THE ANALYSIS OF PROTEINS

There are several different types of techniques that are used in the analysis of several proteins present in living organisms. There are several analytical techniques that are used in the understanding, analyzing and studying the proteome of the living organism. There are different stages of analysis of proteins but eh first and the foremost step in the analysis of the proteins is the separation of a sample that consists of a huge mixture of proteins.

The separation of proteins is generally carried out in a two-dimensional gel electrophoresis technique. The protein molecules in the two-dimensional gel electrophoresis are known to be separated based on the isoelectric point or molecular weight of the proteins.

Initially, the proteins are separated based on their charge and then the gel is turned ninety degrees and the proteins are then separated based on their size or molecular weight. This separation technique is generally carried out in two dimensions because of which this method is called two-dimensional gel electrophoresis.

The sample is generally stained with commissive blue because of which the protein molecules re generally seen as spots on the gel. These spots are then treated with several proteases. These enzymes are known to breakdown protein molecules into several peptides. These peptide molecules that are present as spots on the gel are then subjected to mass spectrometric analysis. In addition to charge and size, there are several other intrinsic parameters that used to separate protein molecules and identify the separated molecules.

There is an important technique which is known to separate molecules based on their mobility in the presence of an electric field and this method is called Field Flow Fractionation (FFF). This is a technique in which the molecule is separated based on mobility. The different types of protein molecules are known to have different size and mass of proteins because of which these proteins are known to have different mobility.

This method is done under several changing circumstances such as electrical, centrifugal, gravitational fields and several others. This technique is known to be extremely essential in determining the different components of the protein mixture. The analysis of several protein samples is known to identify different conformations of proteins and their interactions with different proteins.

There are several advantages of proteomics analysis and the proteome analysis is conducted at different stages. The various steps that are part of the proteomic analysis are explained below.

1.8 STEPS IN PROTEOMIC ANALYSIS

There are several steps that are involved in proteomic analysis and the different stages of proteomic analysis are as follows:

- Purification of proteins
- Separation of Proteins
- Identification of proteins

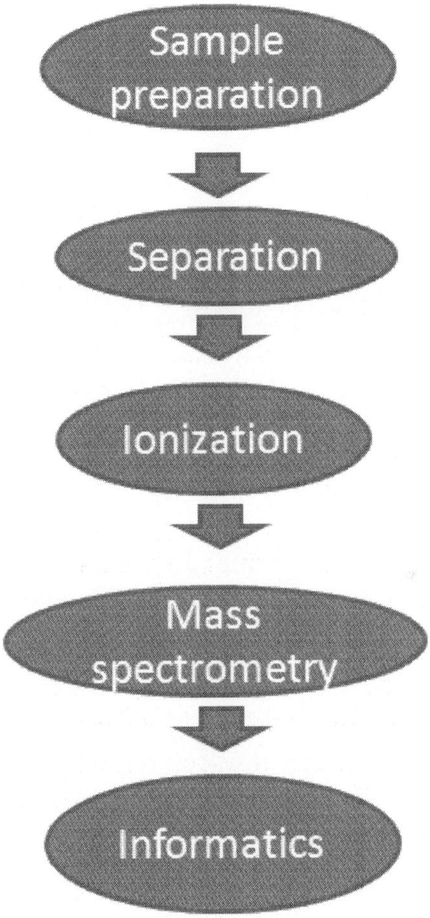

Figure 1.5: Stages in Proteomics.

Source: https://upload.wikimedia.org/wikipedia/commons/f/f3/Flowchart_of_proteomics.gif

1.8.1 Purification of Proteins

This is the first and foremost step that is involved in the study of proteins. This is an important step in the proteomics studies because there are several protein samples that are extracted from the different parts of the cell such as whole-cell, tissue or several other subcellular organelles.

The samples that are extracted from these cell organelles are known to be purified by several procedures such as gradient centrifugation and chromatographic techniques.

The different types of chromatographic techniques that are used are exclusion chromatography, affinity chromatography, and several other techniques.

1.8.2 Separation of Proteins

2D gel electrophoresis is an important technique that is used in the separation of proteins. This method involves the separation of proteins based on their isoelectric points and molecular weights. This separation technique would show several spots of proteins on their gel and these spots are of several proteins that are separated by molecular weight or isoelectric points.

The protein molecules are mostly present as spots because they are used as fluorescent dyes or radioactive probes.

1.8.3 Identification of Proteins

The next stage that is related to the study of proteins is the identification of the proteins. The proteins that are present in the gel are digested by any protease enzyme. The digestion of the proteins by a protease enzyme would lead to the formation of several peptide molecules from these proteins.

These eluted proteins are generally identified with the help of several sophisticated technologies such as mass spectrometry.

1.8.4 Analysis of Proteins

The proteins that are identified and detected are generally analyzed with the help of several sophisticated technologies such as MALDI-TOF (Matrix-Assisted Laser Desorption Ionization-Time of Flight) which is largely based on peptide mass fingerprinting.

1.8.5 Comparative Analysis of the Amino acid

An amino acid sequence that is determined from all these results is generally compared with the available databases that are used to validate several protein sources. There are several online tools that are available for proteomic analysis which is used in the study of proteomics such as Mascot, Popitam, Aldente, Peptide cutter and many more.

Proteomics is known to have several applications in different aspects of life sciences such as practical applications of the life sciences which include the drug development against several diseases. The proteomic technologies are used to make a prediction regarding several diseases, and they are used in the analysis of several normal and they are used for making the protein to gene prediction.

The identification and detection of the protein/gene play an important role in understanding the function of the protein/gene. The techniques that are available in the identification of the gene will play an important role in understanding the function of the gene and it is known to play a vital role in understanding the different areas of disease management and drug development.

Proteomic analysis generally refers to the analysis of the protein from a specific gene sequence. The recent studies in genomics led to the study of the whole genome sequence of several organisms and there are several organisms whose whole genome sequence is available. The availability of the whole genome sequence of an entire organism plays a vital role in understanding the different proteins that are produced by these organisms and the role of these proteins in several biological processes.

There are codes of several genes that are mostly modified at the time of synthesis. There are several types of post-translational modifications that are present that would lead to the functional diversification of these proteins. There are different applications of the proteomics and the various applications of the proteomics must be studied.

1.9 IMPORTANCE OF PROTEOMICS

The active role of the molecules and components of the cell in the survival of living organisms along with their functioning is known to play an important role in scientific studies right from the beginning of biological research. The need to understand the role of the proteins in the functioning of these organisms made it important to give the name "protein" in Berzelius in 1838. The word protein in Greek refers to the word proteins which means "the first rank". The several important roles of proteins in different areas of proteomics research are mentioned below:

A lot of functional information of the genes is characterized by the proteome. The proteome of prokaryotic and eukaryotic cells is known to be different in terms of their complexity and the eukaryotic cells are known to be comparatively complex than the proteome of the prokaryotic cells. The damage range of the eukaryotic cell is known to be extremely high when compared to the prokaryotic cell. The proteins that are expressed by the prokaryotic cells are known to be responsible for the pathogenic mechanisms in different organisms.

The analysis of these prokaryotic proteins is known to be challenging and problematic because of the presence of a lot of diversity in the properties of these proteins. These proteins are known to be having a lot of diversity in their properties such as quantity, molecular size, hydrophilicity, and hydrophobicity.

Proteomics is known to be that area of science which has several technologies that play an important role in the diagnosis and the prognosis of several diseases, and they are known to play an important role in the monitoring of disease. Proteomics is known to be an important branch that is known to play an important role in drug design and drug development.

Proteomics refers to the branch of science which deals with the characterization of the protein, structure and function of proteins, interactions of protein-protein complexes and understand the modifications that occur at any step of expression of a protein.

The proteome of a cell is known to change in several ways with respect to time, cell to cell and other factors. The major reason for this variation in the proteome may be largely due to the external environment in which the living organism is present.

The proteomics of the eukaryotic cells is known to be extremely complex because of several post-translational modifications that may arise at several sites because of several reasons.

Proteomics is known to comprise several latest technologies that are used for understanding the functioning of the gene and this branch of science is known to be a lot more complicated than genomics.

There would be several fluctuations that would be present in the expression of genes. These changes in the expression of the genes can be determined with the help of analyzing the transcriptome or proteome of the cell.

The analysis of the proteome of the cell would then be compared according to the biological states of the cell to understand the differences in the gene expression level. There are several micro-array chips that are being developed for the large-scale analysis of an entire transcriptome of the proteome of the cell.

The major limitation of these microarray technologies is when there is an increase in the synthesis of the mRNA as this increase in the synthesis of the mRNA cannot be measured directly by the microarray.

Proteins are effectors of biological function and their levels are not only dependent on corresponding mRNA levels but also on host translational control and regulation.

Therefore, proteomics is known to be an important area and tool for the characterization of any biological system.

1.10 TECHNIQUES USED AND CLASSIFICATION OF TECHNIQUES THAT ARE USED IN PROTEOMICS

There are different types of techniques and methods that are used in proteomics and the classification of techniques in proteomics is given below and they are as follows:

- Conventional methods in proteomics
- Modern Methods in Proteomics
- High throughput methods in proteomics

1.10.1 Conventional Methods in Proteomics

The conventional methods that are used in the purification of the proteins are largely dependent on chromatography. The different types of chromatography techniques that are used in the purification of the proteins are as follows:

- Size-exclusion Chromatography (SEC)
- Affinity Chromatography (AC)
- Ion-exchange chromatography (IEC)

These methods are largely used for purification of proteins and there are several methods that are used in the analysis of the proteins. The important techniques that are used in the analysis of selective proteins are as follows:

- Enzyme-Linked Immunosorbent assay (ELISA)
- Western blotting

These techniques are widely used for the analysis of several proteins but they lack the capacity to measure the protein expression level. There are some other techniques that are used in the definition and analysis of the protein expression level of proteins.

The different methods that are traditionally used to measure the protein expression level in several proteins are the two-dimensional gel electrophoresis (2-DE) Sodium dodecyl sulfate-polyacrylamide gel electrophoresis (SDS-PAGE), and two-dimensional differential gel electrophoresis (2D-DIGE). These are known to be used in the separation of several complex proteins.

1.10.2 Modern Methods in Proteomics

There are several novel methods and sophisticated technologies that are developed in proteomics. The development of several latest technologies is paving way for the enhanced methods in the analysis of these proteins. There are latest technologies such as protein microarrays or chips that have been made for high throughput and rapid expression analysis of several proteins.

The progress of the protein microarray is known to be extremely challenging because of the properties of certain properties and they are essential for analyzing the function of the complete genome which is extremely challenging.There are several novel technologies that are designed for analysis of several proteins that are being used by mass spectrometry and these methods have been developed to analyze complex protein molecules with high sensitivity. In addition to these technologies, there are certain sequencing methods that were developed to determine the amino-acid sequence of a specific protein.

There are several recently developed techniques that are widely used in the study of proteins and these techniques are used in the quantitative analysis of different proteins that are present in the cell. The different recently developed techniques in proteomics are as follows:

- Isotope-coded affinity tag (ICAT) labeling
- Stable isotope labeling with amino acids in cell culture (SILAC)
- Isobaric tag for relative and absolute quantitation (iTRAQ)

1.10.3 High Throughput Technologies

The recent studies and advances led to the arrival of several high throughput technologies. There are large volumes of proteomics data that is being collected and a lot of bioinformatics databases were established to handle enormous quantities of data for their storage. There are different types of latest bioinformatics tools that are developed for 3D structure prediction, motif analysis, rapid analysis of protein-protein interaction, protein domain and the data analysis of Mass Spectrometry.

The two important high throughput technologies that play an important role in understanding the three-dimensional structure of a protein are X-ray crystallography and nuclear magnetic resonance (NMR) spectroscopy. These are extremely useful in understanding the biological function of any protein.

Several alignment tools have been developed that are useful for the sequence and the structural alignment of peptide sequences. The sequence analysis tools are known to play an important role in understanding the evolutionary relationship of several genes.

The analysis of proteins is known to provide with a lot of data and the proteome analysis is known to provide a complete depiction of structural and functional information of the cell. This information is known to play an important role in understanding the response mechanism of the cell against different types of stress.

These techniques are also important in understanding the mechanism of action of drugs over several single or multiple proteomics techniques. Therefore, there is a need to understand the role of several proteomics techniques and their applications in various areas of science as they are crucial in understanding different aspects.

1.11 APPLICATIONS OF PROTEOMICS

The most common application of proteomics is known to be protein identification. There are several researchers who are using various concepts of proteomics to isolate and display several proteins that are expressed by the genome of the organism. There are several researchers who use proteomic technologies in the isolation and display of several proteins.

Protein identification is known to provide with a lot of immediate information that is responsible for various applications in research. Protein identification is known to provide with a lot of data and analysis regarding the various steps that are involved in the experimentation of proteins.

For instance, the identity of a protein is known to play an important role in determining an anticipated result, validate proteomics methodology that is used in the analysis of compounds, facilitate with a lot of information that is not expected and show a biochemical method which is not working.

A major application of any area of proteomics is the sequences of steps and the experimental approach that is used in the isolation, identification, analysis of compounds and determination of the structure and function of several protein targets.

There are several developments that are generally happening in the field of proteomics and advancements in several technologies related to the Mass Spectrometry. There has been a lot of shift in attention of proteomics from the accurate techniques of proteomics to the methods of data analysis. There is a need for data analysis in the field of proteomics because there is need for acquiring data from a pool of compounds that are not relevant to the identification of proteins and it is essential in consumption of time and effort of proteins.

The major strategy that is key to any proteomics approach is to devise those techniques that are essential for understanding the role of functioning of different low-abundance proteins that are present in a cell. The low abundance proteins in a cell are generally studied by the different techniques related to the differential display of the compounds that are mostly extracted by methods such as affinity chromatography.

In affinity chromatography, there are several affinity columns that are used in designing protein targets. The protein samples are generally exposed to a set of proteases and strict set of washes that are used to remove nonspecific proteins. This strategy generally is useful for the analysis of several proteins that are not essential in the analysis of different set of compounds.

Proteomics is known to be an emerging field in biology which is known to have applications in several areas. The important applications of proteomics are as mentioned below:

- The most important and major application of proteomics is its promising role in understanding, analyzing and designing drugs specifically for a disease. Proteomics is known to play an important role in the development of new drugs for a few diseases.
- The protein expression data is being used for understanding the

role of each protein and it is known to play an important role in designing new drugs or combating for the existing problems in that field.

- There are several novel tools and online data that are available for the development of new drugs and combat complicated diseases.
- Proteomics helps in identifying the competitive molecule that is useful for inactivating various enzyme.
- Proteomics data helps in designing personalized medicine for the patients and would have a significant impact on drug discovery.
- Target identification and validation are the important stages of drug discovery that utilize proteomics data.
- Proteomics involves the study of those molecules that are useful in the discovery of proteins and play an important role in designing drugs.
- The research in proteomics is known to play an important role in designing biomarkers and it enables to understand the effect of these biomarkers on diseases.
- Proteomics is known to assist in screening several compounds that are involved in pre-clinical studies.
- Proteome Mining: Proteome mining is an important field in the area of functional proteomics. Proteome mining is known to be an area of proteomics that is used in the analysis of specific sub proteomes that are belonging to different class of proteins. The principles of proteome mining assume that all drug-like molecules selectively compete with a natural cellular ligand for a binding site on a protein target.
- Yeast Genomics and Proteomics: This is a field of proteomics that deals with the application of the technologies of proteomics along with certain concepts of yeast genetics that plays an important role in delineating the cell signaling events *in vivo*.
- Characterization of Protein Complexes: There are several laboratories that are working on the characterization of protein complexes. Several sophisticated technologies such as Mass Spectrometry are used in the characterization of these proteins.

For instance, Linke et al use multidimensional LC and MS/MS in the characterization and identification of proteins and Msnn and colleagues is known to be another team that is working towards the identification of proteins after the conduction of immunoprecipitation for protein complexes. There are several others in recent times that used the principles of Mass Spectrometry for identification and study of several proteins that are interacting with the Cdc42 effector molecules.

- In recent times, there is a significant gap that is being observed between genetic variant and in establishing causality. The study of the difference between the genetic variant and the original plant is essential for understanding several molecular mechanisms that are present. This study would lead to a better understanding of the clinical phenotype and to identify several novel therapeutic targets.

1.12 PROTEOMICS IN FOOD SCIENCE

Proteome refers to the total set of proteins that are present in a biological sample. The proteome is known to be a lot more dynamic when compared to the static genome and there are several external factors that have an impact on the state of development of the plant, metabolic and biochemical state of the plant, type of the tissue that is present in the plant and several other interactions of the plant. Therefore, the proteome of the plant is known to be an indicator of the several biochemical processes that occur in a living organism.

The proteomic analysis is generally conducted by the gel-based methods and the shotgun methods which are widely used in several areas of proteomics. The steady observation and analysis of proteins is known to play an important role in understanding the biochemical pathways and composition of proteins in several living organisms because of which there is comprehensive knowledge about an organism would be available by proteomic analysis.

This is the major reason for the growth in the use of proteomic analysis in food science and technology. There are different areas of food science in which proteomics is successfully used and the different applications of food science in proteomics are mentioned below:

- Analysis of the food composition of several food products
- Proteomics plays an important role in the identification of food allergens

- Proteomics plays an important role in the safety assessment of several genetically modified food
- There are several marker proteins that are being searched for food authentication with the help of proteomics.
- Proteomics' is widely used for the systemic analysis of the biochemical activity of several food products
- Food processing is an important step in the food technologies and there are several proteomic concepts that are being used to understand the impact of food processing technologies on the different types of proteins that are available in food
- The major application of several proteomic technologies in food science is that these proteomic technologies are used for the improvement of the quality of the food.

Food quality and safety are two major concerns in food science that pose a threat across the world. The safety issues that are present in food science are known to develop large economic losses along with a lot of danger to the health of the consumer.

If there are no efficient methods in the screening of the compounds then there are a lot of negative impacts on the food supply that would affect the confidence of the consumer because of the presence of inappropriate food safety measures. If there are inefficient measures and the lack of proper steps in screening of foods may lead to several losses to the food industry and might create irrevocable losses to the consumer.

There is a need to set up and develop technologies that are reliable and can be used for faster analysis of several food products on the entire food chain. The field of proteomics can play an important role in developing several technologies that can play an important role in the qualitative and quantitative analysis of the food.

There are several developments in the field of proteomics and these developments are gradually improving the focus of proteomics from plant sources of food to animal sources of food. Proteomics is offering promising scope and potential for improvement of quality and security of food with the help of several novel and sophisticated technologies.

The major objective of several researchers is to understand the role and importance of proteomic technologies in improving food quality and safety. There is a lot of research and several studies that are being carried out in application of proteomics in several areas of food processing, allergies and several possible contaminants such as fungi, bacteria and several other pathogenic microorganisms that would affect the sanitary and nutritive quality of the food.

1.13 PERSPECTIVES

The DNA microarray analysis is a technique that enables the analysis of several thousands of genes that can be monitored simultaneously. The presence of such high throughput technologies raises the question regarding the importance of proteomics. There is a need to understand the importance of proteome in several studies as proteins are extremely important for the structure of the protein, produce energy in the protein, they are responsible for protein-protein communication in proteins within the cell, cell signaling, movement of the cells and reproduction in the cells.

Proteins are known to be playing an important role in providing scope for the study of the structural and functional framework for cellular life. The genetic information of a cell is known to be static while the protein complement of the cell is known to be dynamic.

Differential proteomics is known to be a branch of science which is known to play an important role in detecting the proteins that are expressed by different genes. The variations in the levels of the proteins that are expressed by the gene are examined and understood with the help of differential proteomics.

Differential proteomics is known to play an important role in understanding the correlation between several ranges of proteins that are expressed in different cells or tissues. This branch of proteomics is helpful in understanding the onset and progression of a disease.

Proteomics is known to help several people in identification of novel protein markers. The biomarkers that are discovered from proteins are known to play an important role in diagnosis purposes and the study of several molecular targets for the drug discovery of proteins. The protein biomarkers that are helped with different technologies in proteins are known to have several potential applications. The development of several biomarkers is known to play an important role in clinical diagnostic or prognostic purposes.

Biomarkers are known to play an important role in the treatment of an effective optimal therapeutic treatment plan for several groups of patients who are suffering from a disease. The biomarkers are known to play an important role in monitoring the impacts of the treatment that is being provided to different people.

Protein markers are mostly used for increasing the rate and efficacy of several clinical trials. There is a lot of research that is being carried out in the area of proteomic and there is a scope for proteomics to be used as molecular targets for the therapeutic intervention of the disease.

Proteomics is known to be that branch of science which is known to be extremely promising and useful in the discovery, screening, and development of the drugs in several disease conditions. There are several novel technologies of proteins that are being developed with advanced bioinformatics that is being used for the detection of molecular signatures of several diseases that are present in protein pathways and signaling cascades.

The application of several areas of proteomics is known to play an important role in diagnosing the individual proteins or the entire metabolic pathways that are involved in the onset and progression of the disease. It is envisaged that analyzing the cellular circuitry of ongoing molecular networks will become a powerful clinical tool in individualized patient management.

1.14 CONCLUSION

There is a vast amount of information that is being generated by proteomics and genomics research and the genetic information that is generated from the genomics research is known to be completely complementary to the genomics research. Though both these areas are complementary with one another in the information that is obtained from their research, proteomics is known to play an important role in the development of functional genomics.

The study of proteomics and genomics in combination with one another is known to play an important role in the development of a lot of diagnostic and therapeutic products in the future. The role of proteomics in food science and dealing with the problems related to food spoilage and food security is an interesting aspect which is explained and discussed in the upcoming chapters.

REFERENCES

1. .com, b. (2019). *Proteomics and Types of Proteomics*. [online] BrainKart. Available at: https://www.brainkart.com/article/Proteomics-and-Types-of-Proteomics_21302/ [Accessed 11 Jun. 2019].
2. Biology, B. (2019). *Genomics and Proteomics | Boundless Biology*. [online] Courses.lumenlearning.com. Available at: https://courses.lumenlearning.com/boundless-biology/chapter/genomics-and-proteomics/ [Accessed 11 Jun. 2019].
3. C, P., P, R., L, B., A, S. and PM, R. (2019). *Proteomics in food: Quality, safety, microbes, and allergens. - PubMed - NCBI*. [online] Ncbi.nlm.nih.gov. Available at: https://www.ncbi.nlm.nih.gov/pubmed/26603968 [Accessed 11 Jun. 2019].
4. C.S. Cho, W. (2007). *Proteomics Technologies and Challenges*. [online] NCBI. Available at: https://www.ncbi.nlm.nih.gov/pmc/articles/PMC5054093/ [Accessed 11 Jun. 2019].
5. DENSLOW,, N., ELLEN MICHEL, M., D. TEMPLE, M., Y. HSU, C., SAATMAN, K. and L. HAYES, R. (2003). *Application of Proteomics Technology to the Field of Neurotrauma*. [online] Pdfs.semanticscholar.org. Available at: https://pdfs.semanticscholar.org/7b49/49704700feafb690d681d7120f417d848ffd.pdf [Accessed 11 Jun. 2019].
6. MAYR, M. (2019). *The Advantages and Limitations of Proteomics and Lipidomics*. [online] Escardio.org. Available at: https://www.escardio.org/static_file/Escardio/Education-Subspecialty/Courses/Councils/Basic%20science%20summer%20school/Summer%20school%202017%20Presentations/MMayr_Nice_ESC.pdf [Accessed 11 Jun. 2019].
7. Nasri, S. (2019). *Proteomics*. [online] Slideshare.net. Available at: https://www.slideshare.net/sarfaraznasri/proteomics-55931424 [Accessed 11 Jun. 2019].
8. P Horgan, R. and C Kenny, L. (2011). *SAC review 'Omic'technologies: genomics, transcriptomics, proteomics and metabolomics*. [online] Obgyn.onlinelibrary.wiley.com. Available at: https://obgyn.onlinelibrary.wiley.com/doi/pdf/10.1576/toag.13.3.189.27672 [Accessed 11 Jun. 2019].
9. R, P. (2019). *Proteome research in food science. - PubMed - NCBI*. [online] Ncbi.nlm.nih.gov. Available at: https://www.ncbi.nlm.nih.gov/pubmed/19690740 [Accessed 11 Jun. 2019].

10. R. Graves, P. and A. J. Haystead, T. (2002). *Molecular Biologist's Guide to Proteomics*. [online] Microbiol Mol Biol Rev. Available at: https://www.ncbi.nlm.nih.gov/pmc/articles/PMC120780/ [Accessed 11 Jun. 2019].
11. science.ru.nl, V. (2011). *Replication/Transcription/Translation*. [online] Vcbio.science.ru.nl. Available at: https://www.vcbio.science.ru.nl/en/virtuallessons/cellcycle/trans/ [Accessed 11 Jun. 2019].

2

Concepts and Techniques Used in Proteomics

CONTENTS

Proteomics is a widely used technology that is used in study of proteins. Proteins are extremely important molecules which have several applications and they are used for comprehensive understanding of the molecular mechanisms in living organisms. There are several popular techniques that are used in the study of proteins. The principle of these techniques has been explained in detail along with the procedure of these techniques. The different molecular techniques and separation techniques of proteomics have been explained in detail. The latest technologies in proteomics such as mass spectrometry, MALDI-TOF and other methods such as Tandem Mass spectrometry have been explained in detail. There is need to understand the principles and procedures of these techniques to understand their role and importance in several areas related to food science and technology.

2.1 INTRODUCTION TO THE CONCEPTS AND TECHNIQUES IN PROTEOMICS

Proteomics is known to be a concept which largely deals with the analysis of several different types of proteins. Proteomics is an area of biology which deals with the huge scale analysis of several proteins that are present in a living organism at a specific period of time. There are several techniques that are a part of proteomics which deals with separation, segregation, and analysis of proteins.

There are several novel and sophisticated technologies such as mass spectrometry that are used in the analysis of proteins. The systems of several living organisms are known to be extremely complex and there is need for sophisticated technologies to properly analyze the protein samples through different proteomic methods.

The important methods that are traditionally used for the analysis of proteins or peptides are chromatography or electrophoresis. The advancements and developments further in this area led to the arrival of several sophisticated technologies such as mass spectrometry. There are further several advancements in the field of proteomics such as gel electrophoresis to mass spectrometry which is used for comprehensive analysis of proteins or peptides. There are different stages after electrophoresis such as detection of proteins and the process that leads from a gel piece to a solution of peptides.

The proteome is the reflection of a living organism which is extremely dynamic, and a living organism is an extremely dynamic ad it constantly changes the response to different stimuli. Proteomics is an area of biology which not only includes the structural and functional knowledge of proteins, but it is important to study the changes and interactions of several proteins which are important. The study of proteins generally includes the study of intracellular localization and quantification of these proteins according to their abundance.

The recent developments in the field of proteomics show the development of several novel methods such as Mass Spectrometry which is mainly matrix-assisted laser desorption/ionization-time of flight (MALDI-TOF) and electrospray ionization trap (ESI-IT). These tools are known to be extremely important as they are extremely precise and essential in the diagnosis of several important types of proteins.

The history of the proteomics started in the 1970s which indicated the development of two-dimensional gel electrophoresis (2DE). These methods paved way and scope for the displaying of several different thousands of proteins on a single gel (O'Farrell 1975; Klose 1975).

The development of several tools of bioinformatics and proteomics are playing an important role in the high-throughput protein sequencing of proteins and they are essential for the identification, quantification and functional analysis of proteins.

There are several advances and research that is currently available in the field of proteomics which is increasing its application in the food composition, safety, quality and biological activity of proteins. There are is a steady increase in the use of proteomics studies in several areas of food science and technology.

The most commonly utilized workflow of proteins is generally called as the bottom-up approach. In this strategy of proteomics called as bottom-up approach, the proteins of interest are generally converted into peptides by application of several economically important enzymes such as trypsin and the peptides that are formed after the application of these enzymes are mostly analyzed by mass spectrometry. (Aebersold and Mann 2003).

There is another strategy that is used in proteomics which is called top-down proteomics which involves the characterization of several proteins that are intact with the help of a mass spectrometer.

An alternative approach that is used in proteomics is generally called as the top-down proteomics which is used for the characterization of several proteins that are intact with the help of a mass spectrometer. The top-down approach does not include pre-enzymatic digestion of proteins which utilizes several ion fragments that are mostly generated by different MS fragmentation techniques (Keleher 2004).

The top-down proteomics approach is feasible to execute as it is known to deliver high accuracy in the mass estimation of the new high-resolution mass spectrometers.

2.2 BOTTOM-UP PROTEOMICS APPROACH

The most commonly used traditional method of proteomics that is used in the analysis of protein samples is the 2D gel electrophoresis. In this method, the proteins are first separated on a polyacrylamide gel matrix. The proteins are separated on a polyacrylamide gel matrix based on two different properties and the two important properties that are used in the separation of the proteins are as follows:

- Isoelectric point(pI)
- Molecular point (Mr)

The proteins are separated on a gel-based on these two properties and these proteins are later visualized with the help of several staining procedures that are available (Miller et al. 2006). The separated proteins are excised and hydrolyzed with the help of different sets of protein digestion enzymes such as trypsin which are leading to the formation of several peptide fragments for each spot of protein on gel. These peptides are mostly detected and identified with the help of Mass spectrometry.

The gel-based proteomics methods are known to be the most commonly used method in the analysis of proteins. This is the most commonly used method because it allows high protein resolution. The equipment and apparatus that is required for the gel-based approach of proteomics are extremely less because of which this method is the most widely used method.

This method is known to be an extremely important method that is used in the analysis of some nonmodel organisms such as the seafood samples. This is a method which is mostly used in the identification of several protein samples related to nonmodel organisms because these analyses are generally comparative analysis which involves a comparison of different proteins of interest to several orthologous proteins from different species with the help of some *de novo* mass spectrometry sequencing strategies.

There are several limitations for different gel-based methods and an important limitation is that the separation of hydrophobic and poorly soluble proteins along with their sensitivity is mostly limited by the availability of different types of detection methods.

2.2.1 Gel-Free Method

The gel-free method in proteomics is commonly called as the shotgun proteomics approach. There are several proteins that are available as a mixture of proteins which are mostly digested by the addition of enzymes such as proteases. The addition of proteases to a mixture of proteins would digest the peptide chains and lead to a mixture of peptides.

This mixture of peptide chains that are formed is separated by the LC which is mostly the reverse-phase chromatography technique. These peptides are eluted with the help of the reverse phase chromatography and the resultant peptides are analyzed with the help of a mass spectrometer.

The use of chromatographic separation of several peptides in a majority of these mixtures becomes difficult and the maximum potential of these one-dimensional chromatography methods is not enough to identify several different types of individuals' peptides.

Because of this, a multidimensional LC in combination with mass detection is widely used in the analysis and detection of several proteins (Motoyama and Yates 2008). A strong cation exchange resin that is coupled with Reverse Phase Chromatography is another method that is popularly used in proteomics.

The recent developments in the field of proteomics led to the development of several novel technologies such as a gel-free approach for analysis of proteins. The development of this gel-free approach indicated that this method has a significant number of advantages over the 2DE. An important advantage of a gel-free approach over as 2 DE is as follows:

- Gel free method offers higher levels of sensitivity over the 2-dimensional gel electrophoresis.
- Gel free methods offer easier reproducibility as it provides automated procedures that are easy to execute and have an advantage over the other gel-based methods
- This gel-free method is known to have a lesser impact on the intrinsic protein characteristics such as (pI, Mr and many more).

Nevertheless, gel-free approaches have the disadvantages that qualitative and quantitative information on protein isoforms is often lost and that the identification of species whose genome is not fully sequenced is more difficult or not possible.

Proteomics is a field of analysis in biology that is concerned with the complete analysis of entire proteome. This refers to the extraction and analysis of the protein content that is a part of the cell, tissue or a complete living organism. This field started developing nearly thirty years back and from then the characterization of several groups of proteins that are present in the living organism has been the primary objective of this area.

The proteome is known to extremely complex in living organisms because of which there is need for constant research and development of novel methods and techniques in the chromatographic separation and detection of different peptides and proteins.

The important methods that are related to the peptide analysis of proteins also include sample cleanup, fractionation, and preconcentration of protein samples become crucial steps in the identification and detection of proteins. There would be a need for analysis of different proteins from various types of biological samples that are generally required in modern proteomics.

2.3 METHODS OF PROTEIN STUDY

There are three important methods that are generally preferred and used in the separation and analysis of different protein samples and some of these samples are:

- Denaturing polyacrylamide gel electrophoresis (PAGE)
- Sodium dodecyl sulfate polyacrylamide gel electrophoresis (SDS-PAGE)
- Two-dimensional gel electrophoresis
- High-performance liquid chromatography (HPLC)

In addition to these methods, there are two other important methods that are used in the analysis of proteins such as capillary electrophoresis and affinity chromatography. These two methods are some other popular methods that are used in the study and analysis of protein in proteomics.

One of the separation techniques that are extremely popular and widely used in separation and analysis of proteins in proteomic studies is SDS-PAGE. The reason for the popularity and wide application of the SDS-PAGE in the separation of proteins is widely due to the simplicity and reproducibility of these proteins and in addition to all these factors is because of the economic significance and simplicity in the use of this equipment.

The SDS PAGE is a method that was initially explained by Laemmli in 1970 which is used for the separation of the proteins. This method earlier did not use amino acid sequence but it uses primary structure or size as criteria for separation of proteins. This method is used to assess the purity of the samples or to estimate the molecular weights of unknown proteins.

A technique called isoelectric focusing was applied to protein and peptide samples around the same time when SDS-PAGE started. A major result of this was the arrival of the concept of two-dimensional (2-D) gel electrophoresis method which is known to have efficient separation capabilities that can easily interface with different types of immunoblotting techniques.

High-performance liquid chromatography (HPLC) is an important technique that is used in the separation of proteins. This technique is the latest technology that has been developed to understand and analyze different organic molecules and ions. High-performance liquid chromatography (HPLC) refers to a type of separation technique that refers it the analysis of various organic molecules and ions.

The instrumentation that is used in conventional high-performance liquid chromatography technique is known to be similar to the equipment that is used in conventional HPLC instrumentation such as separation columns, pumping systems, and detectors which were initially used for the conventional analysis of proteins and peptides along with their applications in their proteomics research.

2.4 MODERN TECHNOLOGIES IN PROTEOMICS

Mass spectrometry is known to be an important technology in proteomics. The application of principles and techniques of mass spectrometry to the analysis of proteins for qualitative, quantitative and functional analysis of proteins is known from times immemorial. These techniques play an important role in the analysis of the entire proteome of an organism which is derived from several complex mixtures and the analysis of the entire proteome is known to play an important role in understanding the cellular function.

Mass spectrometry refers to an analytical tool that is used for measuring the mass to charge ratio of several molecules that are present in a sample. The values of several mass-to-charge ratios are used in several types of measurement and they are known to play an important role in calculating the molecular weight of several proteins and are used in proteomics to analyze different proteins that are present in the sample. Mass spectrometry plays an important role in the identification of several unknown proteins and helps in the determination of the structure, molecular weight and chemical properties of several protein molecules that are present in the sample.

The most commonly used methods that are used in the proteomic methods are electrospray ionization and matrix-assisted laser desorption/ionization (MALDI). There are several soft ionization methods that are present and these are used to generate low internal energies when the protein molecules are fragmented to form peptides and these are representatives of the soft ionization techniques.

Protein microarrays are known to be those techniques that have progressed at a faster rate for the identification, quantification and functional analysis of the proteins. These are widely used in different areas of proteome research. There are several multiplex assays that have been enabled for a higher precise characterization of the proteins. Several of these techniques are used for the study of different complex protein-protein interactions along with several peptides, low molecular weight compounds, oligosaccharides of DNA and several other molecules.

The next generation of microarrays that are used in proteomics for analysis and detection are known to have the capacity of high-throughput sequencing, low-cost biomarkers, ultrasensitive analysis is used in proteomics. There are several novel techniques that are used in combination of nanotechnology, microfluidic networks, and advanced data analysis tools. This procedure generally increases the process of protein biomarker discovery and characterization of several disease-specific pathways.

2.5 CLASSIFICATION OF TECHNIQUES IN PROTEOMICS

There are several techniques that are present in proteomics and some of the important techniques are mentioned hereunder:

- Molecular Techniques
- Separation Techniques
- Protein Identification Techniques
- Protein Structure Techniques

2.5.1 Molecular Techniques

Some of the important molecular techniques related to proteomics are as follows:

- DNA Microarrays or Gene Chips
- Differential Display
- Northern Blotting/Southern Blotting
- RNAi (small RNA interference)
- Serial Analysis of Gene Expression

Yeast two-hybrid analysis

2.6 SEPARATION TECHNIQUES

Some of the important separation techniques in proteomics are given as follows:

- 1D gel electrophoresis
- 2D gel electrophoresis
- Capillary Electrophoresis
- Chromatography: Some of the important types of chromatography are HPLC, SEC, IEC, RP, Affinity and many more
- Protein Chips (Protein Microarray)

2.6.1 1D Gel Electrophoresis

This method is called as one-dimensional gel electrophoresis which is traditionally used in the analysis of proteins. This method is widely used for the separation and analysis of proteins. This method separates proteins based on molecular weight. The average molecular weight range of proteins that are used in the separation and analysis of the proteins is mostly 10 to 300 kilodaltons (kDa).

In the one-dimensional gel electrophoresis, the samples that contain a mixture of proteins are dissolved in a negatively charged detergent called Sodium Dodecyl Sulfate (SDS). SDS is known to be a negatively charged detergent which has both hydrophilic and hydrophobic regions. Hydrophilic regions are those regions that are not associated with water and the hydrophobic regions are those regions that do not have any affinity with water.

Sodium dodecyl sulfate is a compound that has the capacity to bind to proteins with the hydrophobic region and this detergent can bind with the hydrophilic region. The presence of this sort of protein-SDS-Water interaction will enable to increase the solubility of even insoluble proteins because of which SDS is considered as the best detergent for dissolving several different types of protein mixtures. These proteins are known to be completely denatured from the original state by the SDS

There is an electric field that is applied to the gel and because of the negative charge of the SDS that is present in the gel, the proteins are known to move faster through the clear acrylamide matrix towards the positive electrodes in the gel. This acrylamide matrix is known to have several holes which are used to sieve out the proteins from the gels based on their molecular weight.

The proteins that are small in size move faster in the gel whereas the proteins that are large in size move slowly. Thus, molecular weight is an important concept that is used in the separation of proteins based on their molecular weight. There is a standard with different molecular weights that are included on every gel to allow the detection of the size of the protein.

A number of foods, supplement, dairy, personal care and several other industries are known to use the testing services of 1D gel electrophoresis to test and analyze several proteins in their products. There are majorly two different types of quantification of proteins that are available in relation to 1D gel electrophoresis and they are relative dye-binding capacity of the proteins or percentage by weight in proteins. Among these methods either of the one method is used for the quantification and analysis of proteins.

Relative Dye Binding Method Used in Quantification

This refers to the amount of binding capacity of the proteins that are present in the proteins that are present in the sample.

Percentage by Weight Protein Quantification

This refers to the percentage by weight of every individual protein that is present in the sample and this protein is compared against a standard to understand the percentage by weight component of the protein. In this method, there are two important methods for the quantification of the protein one of them is the molecular weight distribution in the protein that is present in the sample and to conduct a customer analysis of the proteins that are present in the sample.

Working Process of 1 Dimensional Gel Electrophoresis

This method is sometimes called an SDS page which is a method that is widely used for the separation of a combination of proteins based on their size. A food sample is stained with the Coomassie blue stain that is loaded onto a gel. These food samples are run on a gel and they show different protein patterns in the gel and later they are analyzed for several different types of studies.

In the 1 D gel electrophoresis there are several samples that are measured, solubilized and denatured in a specific SDS-containing buffer and then they are separated on a gel matrix. A stain called Coomassie blue is used and the sample is stained with this particular dye and run on a gel on which the protein samples are separated based on isoelectric point of molecular weight.

Protein digestion enzymes such as protease are added to this sample and the proteins present in the gel are fragmented to peptides. These peptide fragments are analyzed further with the use of mass spectrometry.

2.6.2 2D Slab Gel Electrophoresis

This is a biochemical method that is used for separating different types of complex mixtures of proteins that are available. The first step in this process is the separation of proteins based on their charge or isoelectric focus point. The size of the proteins is used for separation of the proteins in sodium dodecyl sulfate (SDS) slab gel electrophoresis.

The starburst pattern of protein spots is visualized on the final 2D slab gel by staining of the proteins or western blotting. The proteins that are separated by 2D slab gel electrophoresis are detected, identified and quantified with the help of mass spectrometry.

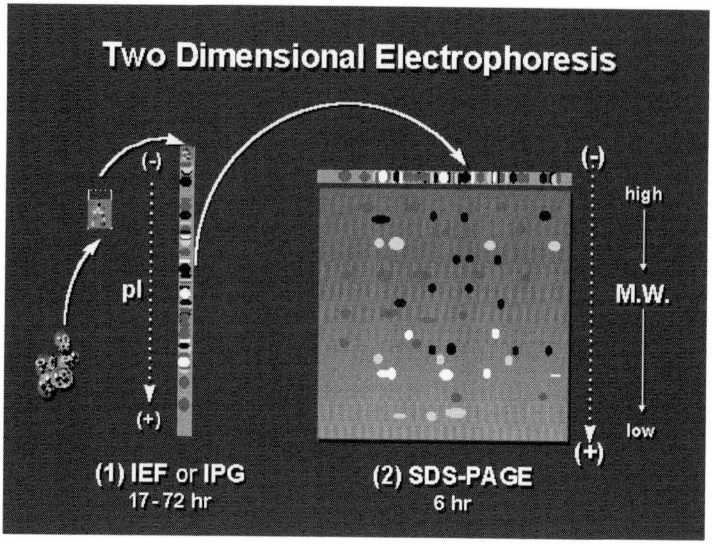

Figure 2.1: Representation of 2D Gel Electrophoresis.

Source: https://upload.wikimedia.org/wikipedia/commons/a/a2/2D_electrophoresis.png

There are two different variations of 2D gel electrophoresis that are used across the world and they are as follows:

- The general classic method that is used for the isoelectric focusing which uses several carrier ampholines that are polymerized in acrylamide tube gels to generate the pH gradient. There is a nonionic detergent that is added to the mixture in this SDS. This detergent is compatible with SDS and this method is called as 2D-SDS PAGE.
- The other method is a method in which the pH gradient is immobilized and it is made commercially available for use to support several IPG stripes. These strips are known to be incompatible with the main detergent SDS because of which it becomes extremely difficult in sample preparation.

2.6.3 Capillary Electrophoresis

Capillary Electrophoresis is a high-resolution separation technique that is used for separation and analysis of several amphoteric compounds known as proteins or peptides. A majority of these analytes or compounds are mostly separated based on pH values in the presence of an electric field. This method is known to be an important method as it is was the first method that was used in the analysis of different peptides.

There were several groups of people that used Capillary electrophoresis in the 1990s for the fractionation of peptides and identification of peptides with the help of Ultraviolet rays or with the help of Mass spectrometry.

This method was initially used for fractionation of the bovine and chicken cytochrome cutis method was initially used for comparative analysis of theoretical pI values and experimentally obtained pI values. There were several limitations that were present with respect to this method and an important limitation is that UV light at 280nm was used in this method. The use of UV light at 280nm will lead to the analysis of only tyrosine and tryptophan-containing peptides. The ultraviolet light at 280nm was used for the detection of several peptides because the ampholyte samples used to generate a pH gradient that is absorbed below the 280nm.

The research studies that were carried out by the Vonguyen et al. nearly a year later published their results similar to the results mentioned above. In addition, the Kasai's research group stated that there are some synthetic peptides that were used as isoelectric point markers which were used to accurately determine the pI of several proteins and peptides by the method of capillary electrophoresis (Shimura et al., 2000).

There are several other methods that have reported the presence of a multidimensional system which is used to separate several tryptic peptides with the help of CIEF that is coupled to nano-HPLC (Kuroda et al., 2005) or capillary RPLC (Chen et al., 2003).

IEF separation of yeast cytosol peptides in fused silica capillaries was performed by Shen et al., and resolved peptides as close as 0.01 pI units (Shen et al, 2000).

The capillary electrophoresis is known to be high-resolution protein separation technique that is widely used for separation and analysis of proteins. There are certain limitations and complexities that are present because of this method because of the presence of certain complicated switching valves and capillary traps which make it extremely difficult for few users who are just beginners. The application of several capillary electrophoresis applications to peptides is known to be extensively reviewed by Kasicka (Kasicka, 2008).

Capillary electrophoresis is an important analytical technique that is used by researchers in different applications of food science and technology.

This method is known for its benefits is analysis several food products such as milk, wheat, egg, meat, fish proteins, and several other proteins and peptides. There are several studies and research that is being carried in the areas of food science to solve several different problems of food science and technology.

The capillary electrophoresis is widely used in a different area of food technology such as assessment and analysis of different problems that are present in the technological process of food technology. This method is also used to test the quality and authenticity of animal foods. There are several future applications of capillary electrophoresis in various areas of food science and technology.

2.7 PROTEIN ARRAYS

The different types of ligand binding assays that are used to measure the protein-protein interaction of the proteins while studying the protein-protein interactions or protein-nucleic acid interactions of several molecules are done with the help of protein arrays.

2.7.1 Types of Protein Arrays

There are different types of protein microarrays and the important categories of microarrays are given below:

- Analytical Microarrays
- Functional Microarrays
- Reverse Phase Microarrays

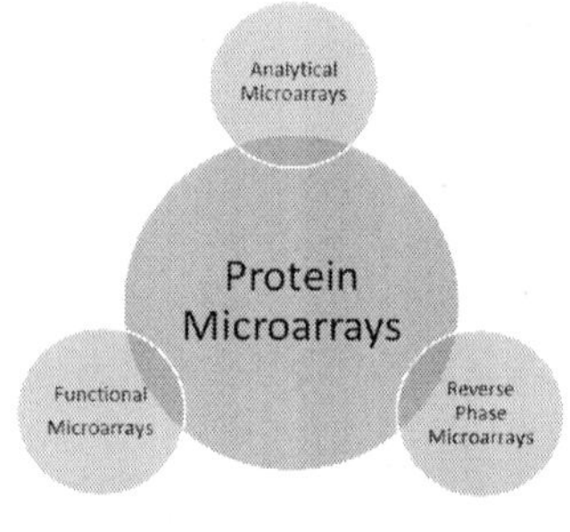

Figure 2.2: Types of Protein Microarrays.

2.7.2 Analytical Microarrays

The analytical microarrays are also called as capture arrays. The analytical microarrays are a type of protein microarrays that are used to understand and analyze the following aspects:

- They are used to understand the expression levels of proteins
- These microarrays are used to measure the binding affinities and specificities
- These microarrays play an important role in understanding the role of these cells to a particular factor
- These play an important in detecting, identification and profiling of several plant tissues.

2.7.3 Functional Protein Microarrays

These protein functional protein arrays are also called as target protein microarrays. This procedure involves the use of several immobilized purified proteins that are used for the identification of different interactions such as protein-protein, DNA, RNA or any other biological processes.

These are used for conducting and understanding the enzymatic activity of different types of proteins that are being analyzed. These methods are different from several analytical methods that involve the identification of various full-length proteins.

2.7.4 Reverse Phase Protein Microarray

This is a type of protein that uses several methods that are used for the determination of different types of altered proteins and other agents that are mostly used for disease.

These are used for the analysis of several parts of post-translational modifications which are mostly altered because of disease.

2.8 PROTEIN IDENTIFICATION TECHNIQUES

The important aspect of any protein identification technique is sequencing. Sequencing refers to the techniques that focus on the identification of the sequence of peptides and amino acids. Sequencing is used for identification of the three-dimensional structure of amino acids and the sequence comparisons among various types of analogous proteins which involve the functioning of the protein and they involve the evolutionary relationship of the proteins.

There may be several diseases that are led to the change of a particular amino acid in the protein and these have an impact on the functioning of the protein. The sequencing methods help in identifying the change in sequence and analyze the reason for the change in the functioning of the protein.

There are several techniques that are used in protein identification and some of the techniques that are used in protein identification are as follows:

- Edman Sequencing
- Microsequencing
- Mass Spectroscopy

2.8.1 Edman Sequencing

The process of purification of a protein by steadily and gradually removing one residue of an amino acid at a given time is known to be Edman Sequencing. The problem of hydrolysis of proteins that arises while studying the proteins is resolved with the help of Edman sequencing. This method is designed by Pehr Edman and this is a widely used method that is used for the labeling of a peptide that is used for cleaving a peptide molecule.

This method was designed with an intention that removal of one sequence at a time would create very little damage to the entire sequence of the protein. This method is known to be extremely helpful in the study of proteins and it is known to play an important role in protein sequencing by leaving back two constituents of the peptide molecule that is used.

This method is continuous to all the amino acid residues that are present in a molecule by removing one residue of the amino acid at a time. This method is a prominent method that is used in the determination of the sequence of the amino acids in the peptide molecule.

Edman sequencing is carried out by denaturing the protein and heating the protein molecule to it and after this process, HCl is added for a long time. This method is effective in separating individual amino acids which are further separated with the help of ion-exchange chromatography.

These are then dyed with the help of the dye ninhydrin that is used in the determination of the amino acid sequence which can be analyzed with the help of an optical absorbance. This method is used in the determination of the composition of the peptide chain but cannot be used for the determination of the sequence of the amino acid.

2.8.2 Mass Spectrometry

There are three important parts of a mass spectrometer (Fenn et al. 1989) and they are as follows:

- Ion Source
- Electrospray (ESI)
- Matrix-assisted laser desorption/ionization (MALDI)

An ion source is generally used to produce ions from a source or sample and there may be one or more mass analyzers such as Q, TOF, IT that aroused to segregate the ions according to mass-to-charge ratio. There is a detector that would be attached to log the number of ions that are coming from the analyzer. This detector is used to generate mass spectra (Canas et al. 2006).

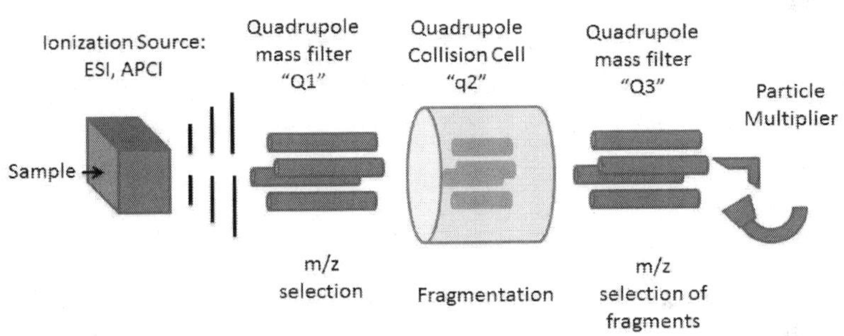

Figure 2.3: Mass Spectrometry Process.

Source: https://upload.wikimedia.org/wikipedia/en/d/d3/Triple_quadrupole_schematic.jpeg

There are two important techniques that were developed to assist mass spectrometry and they are ESI and MALDI. These techniques have universal applications because they are capable of the conversion of the large molecules into gas-phase ions without having an impact on the integrity of the molecules.

This principle of proteomics played an important role in the development and progression of proteomic analysis. These latest techniques and sophisticated technologies played an important role in high-throughput identification of several thousand proteins in one experiment.

2.9 ESI PROCESS

In this process, high voltage may be applied to a solution that contains a mixture of several proteins that are flowing from a narrow capillary tube which produces an electrical spray that is composed of several charged small drops that would be present within the mixture.

These small drops that are produced as a result of the application of high electric field are known to evaporate faster until the number of the charges on the surface becomes very high and they surpass the Rayleigh limit. These drops explode and form several small drops when they are exposed to the electrical field.

The process of application of high voltage electrical field to the solution to analyze ions is repeated until the mixture of these peptide molecules are isolated and are further desolated.

The major advantage of the ESI is that based on the molecular mass and structure the analyte ions would try to acquire multiple charges. The tryptic peptides are those peptide molecules that generally become doubly or triply charged ions.

2.10 MALDI PROCESS

Sample and a dried mixture of the matrix (cyano 4-hydroxycinnamic acid, sinapinic, and others are mixed together. A laser beam of extremely high frequency is fired at this mixture and this energy that comes from the laser is absorbed by several molecules that are present in the matrix. This energy that is absorbed by the molecules is known to lead to desorption of solid crystals.

In the process of MALDI-TOF, the analytes that are present are taken out together along the matrix which results in a dense plum of matrix and analyte molecules that are present in the gas phase. In this gas-phase the proton transfer is known to happen very fast because of which in this process the charged molecules that are formed are known to have only one single charge.

The analyzer is known to be an important part of the mass spectrometer and the ions that are formed are known to be separated according to the m/Z ratio of the ions. The important concepts that are related to the analysis of proteins that are most frequently used in the area of proteomics are Time of flight (TOF) and quadruple (Q). The mass spectrometers are equipment that is mostly constructed with one or two analyzers.

The instruments that are composed of either two or more mass analyzers that are combined together are called tandem mass spectrometers. The TOF is known to consist of a high tube that is present in a high vacuum where the ions, accelerated with equal energies, fly along the tube with different velocities that are inversely proportional to their masses (Weickhardt et al. 1996). The ions are separated based on their m/z values as the ions with higher m/z values are known to reach the detector slower the smaller m/z values.

TOF instruments are generally attached to a source of MALDI and this is called as MALDI-TOF which has been used in the process of peptide fingerprinting for more than twenty years. Quadrupole works by the selective stabilization of ion trajectories through an oscillating radio frequency field created by four parallel rods (Leary and Schmidt 1996).

In this field that is created in quadrupole, the movement of the ions becomes extremely difficult because of which ions with only small and stable trajectories will travel across the quadrupole and will reach the detector.

There are some ion trap instruments such as 3D or linear methods and a major commonality among these methods is that the ions are retained inside a dynamic electric field and these ions are sent into the detector based on their m/z values (Jonscher and Yates 1997).

In the Ion Trap equipment, the process of selection of ions, decomposition and fragment analysis is known to be repeated multiple numbers of times. This process which involves the repeated selection, decomposition of ions along with fragment analysis is called as Msn. These ion trap equipment are known to be high-resolution separation techniques which are coupled to an ESI source.

There are several instruments that are known to offer some high-resolution analysis and these instruments are described below:

Orbitraps and ion cyclotron resonance instruments are known to be coupled to several analyzers to improve their resolution and these analyzers to which the orbitraps and ion cyclotron resonance are transferred are known to increase the efficiency of high-resolution analysis.

In these instruments, detection is based on the production of induced alternating current (AC) with the same frequency as that of the oscillation of the ions inside the instruments. There is a very complex signal that would be produced and it is mostly mathematically processed by the most famous Fourier transform algorithm.

2.11 TANDEM MASS SPECTROMETRY

This technique is a combination of two important steps of mass spectrometry and it plays an important role to get information regarding the sequences of the peptide chains in proteins. The first step in the Tandem Mass Spectrometry is the isolation of the peptide chain.

This peptide chain is fragmented by the application of energy. This energy needed for fragmentation is majorly by two methods and they are as follows:

- Collision-Induced Dissociation (CID)
- Electron-Transfer Dissociation (ETD)

2.11.1 Collision Induced Dissociation

This is a method that is mostly by the collision of the peptide chains with an inert gas to cause fragmentation of this chain.

2.11.2 Electron-Transfer Dissociation

This is a method which causes fragmentation in peptide chains by electron transfer.The fragments of the peptide chain that are obtained after application of energy are used to generate mass Spectra. These mass spectra are obtained to reconstruct the sequence of peptides (Steen and Mann 2004).

2.11.3 Instruments for Tandem Mass Spectrometry

The instruments that contain ESI or MALDI are used along with MALDI-TOF/TOF, ESI-Q-TOF or LIT for analysis of protein by tandem mass spectrometry.

2.12 PROTEIN STRUCTURE TECHNIQUES

There are several types of techniques that are used in the structural assessment of the protein and some of the important protein structure techniques are as follows:

- NMR (Nuclear Magnetic Resonance Spectroscopy)
- X-ray Crystallography
- Computational Prediction

2.12.1 NMR (Nuclear Magnetic Resonance Spectroscopy)

Nuclear magnetic resonance spectroscopy is an important branch of spectroscopy that is used in the determination of the structure of the protein. The nuclear magnetic resonance spectroscopy is a concept of spectroscopy that uses electromagnetic radiation for the determination of the structure of the proteins. The electromagnetic radiation in the radio frequency range of 10MHz to 1GHz is used in this method for the determination of protein structure.

These radio waves are targeted at the atomic nuclei that are kept in the presence of a strong external magnetic field (Bo). The presence of an external magnetic field is known to polarize the nuclei of the atoms in the presence of strong radio waves because of which the atoms are taken into the higher energy state based on the nuclear spins of the compounds.

The removal of the strong external magnetic field will make the spins of the compounds to return to a lower energy state. This arrival of the compounds from higher energy state to lower energy state is known to produce a small quantity of radiation.

This return of the spins of the atoms is called spin relaxation within a small electromagnetic field. The radiation emitted induces a radio frequency signal in a detector coil within the NMR spectrometer, which is then amplified and Fourier transformed to produce the NMR signal (the "chemical shift').

2.12.2 X-ray Crystallography

X-ray crystallography is a form of extremely high-resolution microscopy that is used for the atomic level visualization of several protein structures. Similar to different forms of microscopy and spectroscopy-ray crystallography works based on the utilization of electromagnetic radiation. The electromagnetic radiation is generally used for the analysis of the structure of a protein that is of interest.

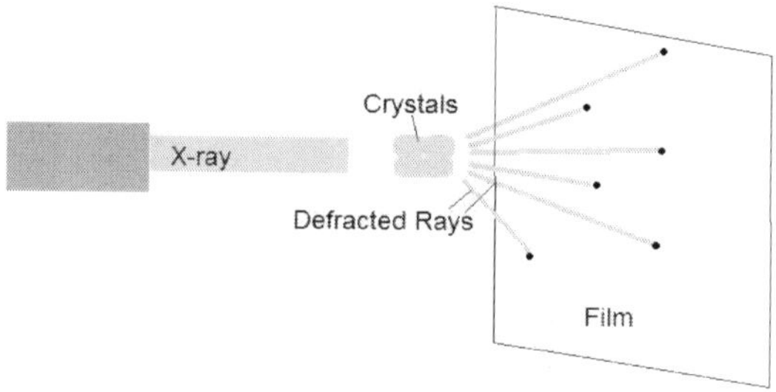

Figure 2.4: Representation of X-Ray Crystallography.

Source: https://upload.wikimedia.org/wikipedia/commons/1/11/X-ray_Crystallography.jpg

The x-ray crystallography is known to be similar to other types of structural determination of proteins because of which it uses high-energy x-rays (λ ~ 0.1 nm). The wavelength of the electromagnetic radiation and the electron that are present in the atom are known to be of same radio wavelength because of which the radiation that is applied on the structure of the proteins is known to get scattered by the electron cloud.

The electrons that are scattered are known to surround the structure of the protein molecule that is present. The beams of electrons are deflected in different directions and these beams are known to constructively or destructively contribute to the production of a scattered molecule which is directly proportional to the scattering angle.

2.12.3 Computational Prediction

There are different computational software's coupled with these experimental methods to determine the structural and functional analysis of these proteins.

The use of this computational software's in the study of proteins is called a computational prediction of proteins.

2.13 CONCLUSION

The two important concepts that were extremely important right from the beginning of proteomics and are especially important in several areas related to proteomics are electrophoresis and 2D gels. The area of proteomics developed along with several protein analysis techniques, the arrival of sequencing technologies such as Edman Sequencing and mass spectrometry.

These techniques of the proteomics are used in the analysis of minute quantities of proteins. Mass spectroscopy is the latest technology that is used to analyze micrograms of protein which is later separated by electrophoresis.

The development and progress in the different areas of proteomics led to the arrival of several techniques that are not dependent on gel electrophoresis and there are several applications of these techniques (Washburn et al., 2001). The development of these novel techniques that are not dependent on gel electrophoresis provided scope to understand and analyze better the strengths and weaknesses of gel electrophoresis.

For instance, the electrophoresis techniques by 2D gels are known to be an approach which is not suitable for the wide-range applications and studies that are involved in proteomics but it is known to provide a better ability to screen and better robustness in the system. These methods are known to improve the efficiency of mass spectrometers by reducing the workload of the mass spectrometers.

The arising and upcoming trends in the field of proteomics indicate that the use of an adequate model is found to be helpful in the analysis of protein and it is a fact that under certain given conditions electrophoresis is known to perform better than chromatograph. There is an absolute need for the combination of RP-nano-HPLC, which is needed for the analysis of the Mass Spectrometry.

There is mostly a comparative approach that is used in protein separation and analysis and this comparative approach states that there is a need for the power of gels to interface different types of protein separation techniques.

The protein separation techniques are known to play an important role in protein microanalysis and these techniques are not completely explained or mentioned prior to the arrival of several gel-free approaches. Gel free approaches are gradually gaining popularity and they are known to be more important than several gel-based methods. The gel-free methods are known to be extremely useful in the analysis of proteins and their quantification. They are gradually becoming the most widely used methods for the analysis of proteins.

There may be several advantages and disadvantages in different types of proteomics techniques, despite the presence of several proteomics methods that are discussed above there are several shortcomings and disadvantage that are present while conducting these methods and the major drawbacks with respect to proteomics are mentioned below:

There is a need to store and process big data files that are related to mass spectrometry

- The acquisition times related to the mass spectrometry are extremely lengthy
- To generate appropriate and accurate data by mass spectrometry it is needed to use extremely expensive, mass spectrometry equipment to increase the efficiency of peptide chain detection and identification
- If the proteins are rare, unique and are not abundant then large dynamic analysis of this protein is generally not compatible even with sophisticated and expensive instruments related to mass spectrometry.

These drawbacks raised different types of question for people and there were doubts and speculations regarding the capacity of the Mass spectrometry related shotgun proteomics. The important speculation was that whether the efficiency of the mass spectrometry-based proteomics in determining the wide-range analysis of the entire proteome with the help of current technologies. (Domon and Aebersold, 2006; Nielsen et al., 2006; Cox and Mann, 2007).

The Nielsen et al have considered that an important strategy for the detection and identification of the low-abundance proteins is to conduct wide range segregation and separation of proteins completely before analysis is a sustainable alternative to the shotgun strategy in protein identification and separation (Nielsen, 2006).

The current observations and studies indicate that among the several techniques that are present in proteomics the 2DE-MS is a technique that is extremely evolved and useful. Figeys et al group states that the 2DE-MS technique is known to be a novel and most widely used proteomics technique that is used in analysis of proteins and they state that " 2DE-MS is now routinely used, while gel-free proteomics experiments are not and, conversely, are increasing in complexity" (Lambert et al. , 2005).

There is a decision tree that is drawn based on the different separation techniques and then interfaces which are widely used in the analysis of proteins. The decision tree helps in identification and selection of a proteomics technique that is widely used and is efficient in the analysis of several techniques that are used in proteomics.

There are several electrophoresis techniques that are widely used in proteomics with various optimum benefits. The decision tree may look complicated but it is actually extremely flexible than it appears to be. There may be certain types of experimental techniques and designs that are used in quantification of methods and this quantification is mostly done with the help of several stable isotopes and in certain cases, a cartographic scheme is more helpful than using a comparative analysis in the study of proteins.

These general schemas are traditionally used for further improvements in the analysis of several techniques and will lead to several developments in the areas of proteomics. The electrophoresis methods and several other methods are known to have an important role in proteomics and their scope, the potential is known to improve to the proteomics techniques.

Thus, electrophoresis separations are clearly a major tool in proteomics strategies, and their use will probably further expand in the near future.

REFERENCES

1. .com, k. (2019). *Overview of 2D electrophoresis Method and SDS Sample Prep*. [online] Kendricklabs.com. Available at: https://kendricklabs.com/2d-overview/ [Accessed 6 Jun. 2019].
2. Meštrović, D. (2019). *Proteomics Methods*. [online] News-Medical.net. Available at: https://www.news-medical.net/life-sciences/Proteomics-Methods.aspx [Accessed 6 Jun. 2019].
3. Piñeiro, C., Carrera, M., Martinez, I., Cañas, B. and Lekube, X. (2019). *Proteomics and Food Analysis: Principles, Techniques, and Applications*. [online] www.researchgate.net. Available at: https://www.researchgate.net/publication/269688154_Proteomics_and_Food_Analysis_Principles_Techniques_and_Applications/download [Accessed 6 Jun. 2019].
4. Poorin, N. (2019). *Techniques in proteomics*. [online] Slideshare.net. Available at: https://www.slideshare.net/NaghmehPoorinmohamma/techniques-in-proteomics [Accessed 6 Jun. 2019].
5. Rabilloud, T., R.Vaezzadeh, A., Chevalle, M., Potier, N., Lelong, C. and Leize-Wagner, E. (2019). *Power and limitations of electrophoretic separations in proteomics strategies*. [online] Arxiv.org. Available at: https://arxiv.org/ftp/arxiv/papers/0909/0909.4158.pdf [Accessed 6 Jun. 2019].
6. Recio, I., Ramos, M. and López-Fandiño, R. (2019). *Capillary electrophoresis for the analysis of food proteins of animal origin*. [online] onlinelibrary.wiley.com/. Available at: https://onlinelibrary.wiley.com/doi/abs/10.1002/1522-2683(200105)22:8%3C1489::AID-ELPS1489%3E3.0.CO%3B2-G [Accessed 6 Jun. 2019].
7. .libretexts.org, c. (2019). *26.6 Peptide Sequencing: The Edman Degradation*. [online] Chemistry LibreTexts. Available at: https://chem.libretexts.org/Bookshelves/Organic_Chemistry/Map%3A_Organic_Chemistry_(McMurry)/Chapter_26%3A_Biomolecules%3A_Amino_Acids%2C_Peptides%2C_and_Proteins/26.06_Peptide_Sequencing%3A_The_Edman_Degradation [Accessed 11 Jun. 2019].
8. libretexts.org, p. (2019). *NMR Spectroscopy vs. X-ray Crystallography - Physics LibreTexts*. [online] Phys.libretexts.org. Available at: https://phys.libretexts.org/Courses/University_of_California_Davis/UCD%3A_Biophysics_200A_-_Current_Techniques_in_Biophysics/NMR_Spectroscopy_vs._X-ray_Crystallography [Accessed 11 Jun.

2019].

9. Reference.md. (2019). *Protein Array Analysis (definition)*. [online] Available at: http://www.reference.md/files/D040/mD040081.html [Accessed 11 Jun. 2019].

3

Proteomics and Plant Food Sources

CONTENTS

Proteomics is basically defined as the study of proteomes. Nowadays, it is a leading technology. Proteomics in the plant plays a vital role in determining the functions, dynamics and interactions of proteins present in a specific plant. The chapter initially deals with the basic introduction to the need of proteomics for plants. The chapter describes plant proteomics as a potential method of food security. Later, the potential of proteomics as a biotechnology tool for the improvement of crop has been explained. In the next part, isolation and characterization of plant protein complexes along with the proteomic evaluation of genetically modified (GM) crops has been briefly described. Certain approaches and challenges in the field of crop plant proteomics have also been explained.

3.1 INTRODUCTION

In a study, it was revealed that on the planet there are close to 925 million people who tend to survive in a hunger-stricken state through their lives. What is even more disappointing is that there may be an addition of close to 2 billion people to this number as the people near the 2050-year mark.

In order to remove the situation of hunger and starvation from the earth, it is required that the people increase the production of food by many folds and improve its supply on similar grounds. This can be done by the combination of various elements and improvising and improving the breeding tools used for plants so that the state of crops can be improved.

The biggest obstacle that the farmers and cultivators face while conducting the breeding programs or the crop improvement programs, comes in the form of a diminished pool of species of the domesticated crops. It is important to identify the genes that have possible uses.

Figure 3.1: Production of food must be increased to fight with hunger.

Source: https://pixabay.com/images/search/maize/

The newly found genes in such a form can go on to be a privilege for the humans if they are placed into the species of the crop that is required and then they are used for the purpose of breeding.

There are several plant sequencing projects are going on right now or have been completed, which have come to the point where they are capable of providing the way of action to the breeders from the modern world so that they are able to work on a large number of genes.

Though there is a limitation stating that the benefits coming from such methods and plan of action can be reaped in the case when the expressions regarding the spatial and temporal form of genes and their activities and the interactions between the products made from them are properly defined.

To state this thing in understandable words, the information regarding the genetic composition of a plat or its DNA is comprehended using a step in between that involves the transcription (mRNA) into a protein.

There are various modern techniques and methods to study the genes expressions that may be spread and found globally and the mechanisms of their activities which involve the understanding of the proteins translated from those genes and that branch of scientific study is commonly known as proteomics.

The comprehension of the word "proteome" can be done from its origin words which state that it is PROTEins expressed by a genOME. In the same way as genomics, the branch proteomics identifies with the comprehension and definition of the whole set of proteins, that are present there at a particular time in a cell.

The genomic studies assist the researchers and scientists in understanding the theoretical aspects of the cellular processes. However, the proteomic studies help them in uncovering the various aspects and elements that help them function and carry out the processes at the cellular level. A genome is static in nature while the proteome possesses various dynamic capabilities.

By understanding the structures and functions of proteins, the various biological activities and their origins can be understood as the study makes the people aware of post-translational modifications (PTMs).

The projects involving the genome sequencing and the research papers based on it are not enough to understand the functioning and the roles of PTMs that they play amid the development and growth of a plant and the response they have to the several conditions of stress and pressure.

The genomics is aided by the branch of proteomics in a way that it completes the whole process of study and understanding the functions. In addition to this, as the branch of proteomics is advancing, it provides the other branches of study with the innovative tools of bioinformatics and computational science.

Moreover, the data achieved from the physiological methods are helping in creating new approaches that can be adopted for the improvement of crops and its study by the methods of networks that are metabolic, regulatory and signaling in nature and form the basis for plant phenotypes.

In a way, the same as the proteomics, the field of biotechnology has gained quite a good amount of popularity in the years that have gone by as it has developed over those years. In the present times, it is a completely developed branch of science and remains as the fastest developing field that deals in the improvement of crops around the world by developing new technologies.

Through the use of biotechnology, the plant breeders get the new capabilities through which they can achieve the goals that they used to think are impossible to achieve, by using the traditional methods of plant breeding. Across the world, the crops that are modified genetically, are grown in the field to fulfill the commercial demands of the people.

To cater to this, the area meant for the cultivation of biotech crop had gone up from 1.7 million hectares in 1996 to 160 million hectares in 2011. Dixon in 2005 mentioned that "Genomics (originally DNA- and transcript-based, but recently extended to integrate the proteome and metabolome) would play a major role in driving plant biotechnology", which indicates that he expected the mentioned pattern to go this way only.

The proteomics has a lot of scope in the field of genetic improvements which concerns the crops for food and biofuel. It also takes into account the quality of food, its safety, the nutritional value it provides, vaccines that may be important for the plants, fungicides based on the study of proteomics and the tolerance of crops to the various biotic and abiotic hurdles and stresses.

3.2 PLANT PROTEOMICS: A POTENTIAL APPROACH FOR FOOD SECURITY

The size of the genome is gradually increasing but the number of genes is not increasing at a proportionate rate. This exciting nugget shows the fact that plants might be involving in combinatorial dealings between genes and gene products, the proteins, to acquire cellular complexity.

Living cells are very complex and may comprise more than 100,000 protein species at any particular time with distinct physical as well as functional properties. The link between proteome as well as phenome is at the center of biological processes.

The field of proteomics comprises the determination of the proteome by using large-scale identification and elucidation of protein and their functions. Proteins are not only required in metabolic activities but they also play a major role in the combination of internal and external signals.

They are exposed to a continuous turnover sustaining cellular homeostasis, which is a vital feature of their regulation. The misregulation of protein expression results in an imbalance in cell milieu. This eventually affects the growth and development of a plant, and decreases crop yield.

Plants mainly contribute to the economic security and sustainability, as they are the key source of food, fiber, fuel and feed. There are countries that have vast plant biodiversity. For example, India, whose agricultural sectors contributes to 8% of the total world's agricultural gross domestic product. India supports 18% of the global population.

On the other hand, almost 80% of the landmass if extremely exposed to external threats. Thus, acceleration of plant productivity, conservation of biodiversity and nutritional security are of supreme significance.

3.2.1 Designer Crops

At present, plant breeders along with the plant biologists are concentrating on the development of trendy crops that are better prepared to survive a broad range of climatic inconsistency and have better availability of nutrients. The approaches of conventional breeding are disadvantaged because breeders need accurate gene modifications with the targeted traits.

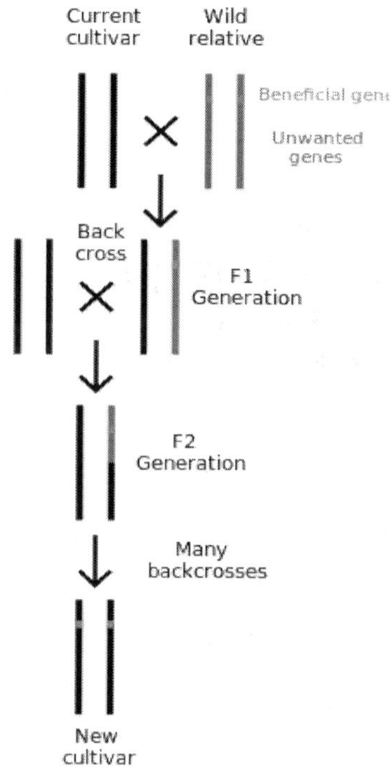

Figure 3.2: Conventional breeding.

Source: https://commons.wikimedia.org/wiki/File:Breeding_transgenesis_cis-genesis.svg

In spite of having scientific potential and infrastructure, India was not a part of the universal human genome-sequencing project. Until now, plant proteomics researchers and scientists were researching the complex biological processes describing the localization of protein species and clarifying the functionality and modifications of the protein complexes.

These days, many plant proteomics laboratories are utilizing high-resolution mass spectrometers for the efficient analysis of proteomes at the cell or tissue or organ level.

3.2.2 Targeted Proteomics

Targeted proteomics should be implemented so as to legalize potentially significant biomarkers for the plant research by using selected reaction monitoring (SRM), parallel reaction monitoring (PRM) or multiple reaction monitoring (MRM). The community of proteomics must aim to offer sets of peptide markers that are spatially unique to provide a snap of Spatio-temporal differential proteome.

Considerable genome annotations and novel genome sequence information would help in adapting various marker peptides for the extensive applicability. In addition, nano-proteomics when combined with the single-cell and micro-dissection analysis could clean the information enormously and provide innovative insights.

A system biology method where a combination of proteomics with phenomics and genomics would improve the meaning as well as the quality of the resulting biological information that requires to be taken. During the past ten years, a wide range of techniques related to protein chemistry has produced large amounts of knowledge on the molecular properties and functions of the individual proteins.

3.3 POTENTIAL OF PROTEOMICS AS A BIOTECHNOLOGY TOOL IN CROP IMPROVEMENT PROGRAMS

3.3.1 Molecular Markers Are to Assist Plant Breeders

The field of proteomics usually provides the identification of novel genes to the plant breeders as well as biologists. The technique of marker-assisted selection (MAS) is the use of DNA markers in the breeding program of plants. MAS has been greatly used in the selection of desired genes for the development of the somewhat higher breeding line.

In the year 1994, Damerval et al. utilized a technique that brought proteomic and components of MAS together. They recognized the protein quantity loci (PQL) that further demonstrated certain spot intensity variation.

Out of 72 analyzed proteins, a total of 70 PQLs were recognized for 42 proteins. Out of these, 20 had more than one PQL. This kind of approach is specifically applicable to different breeding programs.

3.3.2 Charting the Proteome and Its Interaction Map Is Important for Creating a Knowledge Base

Almost all of the cultivated lands come under sub-optimal situations for commercial agriculture. Due to such optimal conditions, around 70% of the crop productions could be lost. To increase the productivity of the crop, proteins and genes that are mainly responsible for disease resistance and stress tolerance have to be continuously recognized.

In this way, a snap of the cellular proteome map at a specific time and under specific conditions enables the recognition of changes in the protein expression. Developments in the MS-based proteomics platforms are considered to be 'New Genomics' as the technique of MS has become a vital tool for the investigation of the PTMs to proteins as well as protein interactions. Such data provides a unique vision on how cells make decisions and are, therefore, the basis of systems biology (Cox and Mann, 2007).

3.3.3 Proteomics Help the Investigations of Abiotic and Biotic Stress Tolerance Mechanisms

The crop plants have to deal with several biotic as well as abiotic stress conditions. As compared to greenhouse nurseries, plants in the agricultural field experience a great combination of different biotic and abiotic stresses.

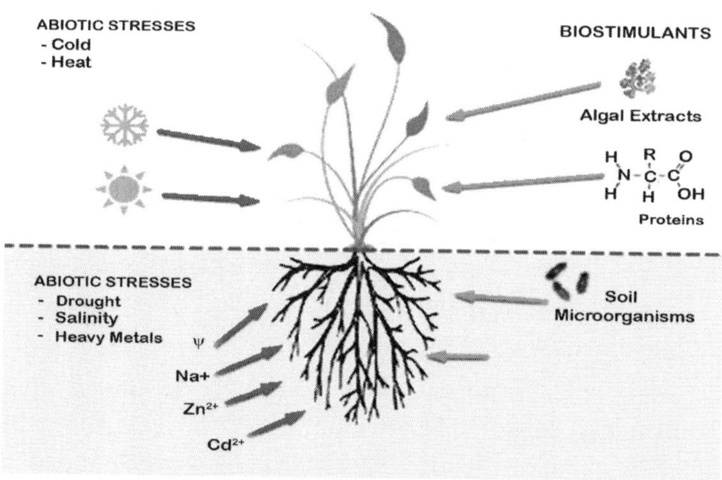

Figure 3.3: Some of the abiotic stress factors.

Source: https://commons.wikimedia.org/wiki/File:Graphical-abstract-Biostimulants-have-protective-effects-against-abiotic-stress.gif

A current estimate recommended that the increase in temperatures from last twenty years have caused a loss of around $5 billion by influencing the production of major crops such as rice, wheat, soybeans, maize, etc.

A plant breeder requires activating the effective protection systems to facilitate the survival of the plant's cells under heat stress conditions. Heat stress tolerance is known as a complex mechanism, which is controlled, by numerous proteins and genes comprising a great number of biochemical as well as physiological changes in the cell.

Global proteomic profiling projects are considered beneficial techniques for increasing the knowledge base of the plant breeders. For instance, a study comparing different wheat cultivars with different heat tolerance capabilities showed that low molecular weight (around 16-17 kDa) heat shock proteins essential for the heat tolerance phenotype.

Proteins from the family of heat shock proteins (HSP) along with the transcription factors upstream of such HSPs have been found to have significant roles in providing the thermotolerance to the crops. Nevertheless, the studies on protein-protein interactions have proved that HSP90 interacts with calmodulin-binding protein (CBP).

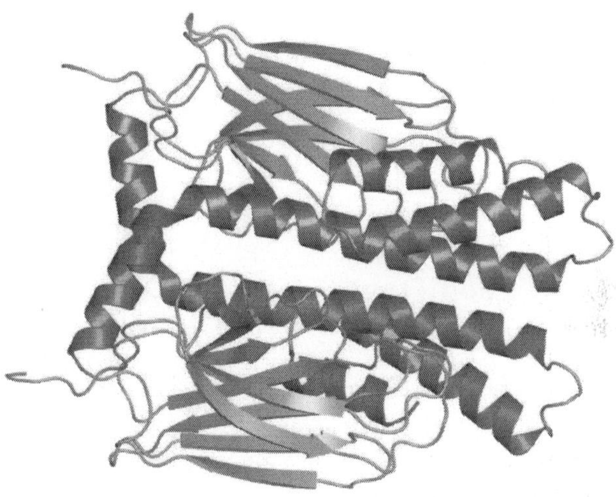

Figure 3.4: Heat shock protein (Hsp70).

Source: https://www.cherrybiotech.com/scientific-note/heat-shock-protein-hsp70

Few proteins other than HSPs, for example, CBP have been recognized in some other proteomics studies in the form of differentially expressed proteins in the heat stress conditions. With this information, the knowledge related to the systemic response of the plants remains limited during heat stress conditions because plant response to single stress is much different from the combination of multiple stresses.

Proteomic approaches are found to be very effective in the study of the molecular mechanism that is involved in the interaction between a plant and its pathogens.

3.3.4 Manufacturing Plant-Based Vaccines Is a Possibility in Near Future

An antigen of interest overexpressed in the plant tissues by a biotechnological method is considered as a plant-based vaccine. In certain situations, dealing with a poorly categorized pathogen, a genomic or proteomic approach is mainly valuable in the identification of the antigens that have favorable features.

A key advantage of the plant-based vaccines is "no safety concerns" (Tacket et al., 1998a, 2000; Kapusta et al., 1999). The vaccine antigens production in plants can be attained through either a stable expression or transient expression systems.

The constant genetic transformation produces a genetically engineered plant that eventually produce the antigens and the plant can be propagated either sexually by seeds or asexually by stem cuttings.

3.3.5 Analyses of Food Quality, Safety, and Nutritional Values Are More Meaningful

Proteomics has been widely used to examine the differences between the nutritional values of the crops by doing the analysis of their proteomes. Iwahashi and Hosoda (2000) stated that it is observed that heat stress increased the invertase expression in tomato and hence, increasing the content of sucrose lead to the production of much sweeter tomatoes.

In addition, proteomics has aided in the proteome map construction by examining the level of protein modifications during the process of barley malting and identifying the proteins linked with the quality of beer.

3.3.6 Biofuel Crop Science is on Right Track to Get Benefitted by Proteomics

Biofuels are mainly produced from the biomass of plants and are known to have the capacity to substitute fossil fuels for the requirements of sustainable bioenergy. Applications of biofuels make a constant balance between the consumption and generation of carbon dioxide in the atmosphere. Unlike fossil fuels, biofuels are made up of pure renewable resources such as plants or photoautotrophic microbes.

The evolution from the use of fossil fuels to the renewable sources of energy is a very complex process that includes scientific as well as socio-economic issues. It is very significant to shift from utilizing first-generation biofuel crops such as corn or sugarcane to the second-generation biofuel crops such as Cordgrass and *Miscanthus* for the production of bioethanol by using lignocellulosic materials found in the plant. And eventually to make a change to the most advanced third-generation biofuel organisms such as microalgae and photoautotroph microbes.

At present, rapeseed, maize and sugarcane are amongst major crops that are being widely used for biofuels. One effective example of adding a novel biofuel crop to the list that is under examination is African grain sorghum. It has widely been utilized as food as well as feed and it is acquiring a lot of attention as an energy crop.

Figure 3.5: Biofuels are gradually replacing fossil fuels.

Source: https://newvitruvian.com/explore/recycle-clipart-biofuel/

3.4 RECENT INNOVATIVE TECHNOLOGIES USED IN QUANTITATIVE PLANT PROTEOMICS

Due to the innovations or developments in the MS-based technologies and quantitative methods, the field of proteomics has evolved as a complementary method to some other approaches such as metabolomics and transcriptomics in the post generic era.

Proteomic analysis is attained through:

- Separation and identification of proteins based on 2-DE or coupled gel-free shotgun liquid chromatography tandem mass spectrometry (LC–MS/MS) platforms.
- Explanation of functions of proteins and their functional networks in the plant signaling and metabolic pathways by doing the analysis of protein mapping, classification of PTMs as well as protein-protein interactions.
- Strategies of bioinformatics and the utilization of databases for the model as well as non-model plant species have been used in the comparative and descriptive proteomic studies of the plant development.

Developments in the technique of liquid chromatography-based separation as well as label-free quantitative proteomic analysis of a great number of proteins that are obtained from complex samples of plants have currently been discussed.

However, modern gel-free proteomics approaches are more developed and they can easily provide detailed information on relative changes in the expression of proteins as compared to 2-DE gel-based methods thathave limitations.

The development and use of label-free quantitative techniques of analysis along with 2-DE gel was the novel realization that actually helped to resolve the major drawback of gel-based separation, quantitative analysis of proteins linked with co-localization or co-focusing of various proteins in one single spot or band on the gel.

Wheat is one of the most efficient agricultural crops. On the other hand, until now, large-scale proteomic studies were very hard to carry out because of the lack of available genomic information to facilitate identification of proteins.

At present, targeted MS-based quantitative methods such as selective reaction monitoring (SRM) have been established for the identification of definite proteins with causative functions in agronomically significant traits (Jacoby et al., 2013b).

The SRM method is completely based on highly precise detection and earlier it was specifically based on triple quadrupole MS platforms because of the requirements of two distinct stages of mass filters. The benefits and disadvantages of this application have been studied and experimentally demonstrated elsewhere.

The major aim of the SRM approach in the field of plant proteomics is biomarker validation in crops. This further allows the phase of discovery with more quantitative and qualitative comparative studies focusing on finding potential candidates that are effective in the stress responses.

Such highly sensitive and selective quantitative approach can be very effective for the biomarker validation and for the development of novel assessment methods of stress tolerance that will enable the identification of genotypes, and eventually the development of gene targets for marker-assisted breeding.

Another approach has been developed recently for label-free shotgun proteomics (MS^E) centered on data-independent acquisition protocols that have a prospective to recognize peptides from complex samples in a consistent, rapid, and sensitive manner, and provides a higher dynamic range for peptide quantification (Buts et al., 2014).

In this way, ultra-performance liquid chromatography is applied which is united to an LC-MS/MS run where a changing level of energy allows to get accurate or precise precursor masses at very low energy and to take fragmentation spectra of all of the parent masses at very high collision energy in only one analytical run.

Although, this approach is used at a very initial stage of development and hence has various challenges regarding the issues in the interpretation of quite complex composite fragmentation spectra causing poor protein identification rate.

Currently, parallel data-dependent runs are required to attain all needed information, maintain the databases of the individual peptide fragmentation spectra and then link them to the MS^E (Buts et al., 2014). Such kind of approach was successfully applied in the quantitative analysis of allergenic proteins in the extracts of wheat grain, in the recognition of glutenins and gliadins in wheat grain and quantitation of proteins linked with celiac disease and baker's asthma (Uvackova et al., 2013a, b).

An alternate strategy that connects very high specificity data-independent method of acquisition with a new targeted data extraction approach to obtain the resulting fragment ion data sets has been verified in recent times (Gillet et al., 2012).

This technique is termed as SWATH-MS. It is completely based on sequential time-as well as mass-segmented acquisition, which eventually generates fragment ion spectra of all precursors in two user-defined dimensions, retention time and *m/z* space, resulting in complex fragment ion maps.

The analysis of multiplexed data sets needed the development of different types of data analysis strategy that utilizes previously attained information available in the spectral libraries to obtain the fragment ion maps for targeted extraction as well as quantitation of particular peptides of interest.

The consistency and accuracy of SWATH MS were established to be analogous to the SRM approach (Gillet et al., 2012). One of the key benefits of the former, improving most restrains of the present methods of proteomics, is the iterative reflective re-mining of the attained data sets for targeted extraction.

This type of approach provides extraordinary possibilities for the qualitative as well as quantitative profiling not only in the field of proteomics but also in the field of lipidomics and metabolomics.

One of the key limitations for the development of proteomics is the absence of effective bioinformatics tools with new algorithmic solutions to the processing of MS data that are lagging behind the significant developments taking place in the protocols and instrumentation (Cappadona et al., 2012; Smith et al., 2014).

3.5 ISOLATION AND CHARACTERIZATION OF PLANT PROTEIN COMPLEXES BY MASS SPECTROMETRY

As various researches have concluded that high-resolution Liquid chromatography mass spectrometry (LC-MS) strategies have dealt with some of the problems with direct analysis of quite complex peptide solutions, other issues have also arose from metabolites and proteases in high profusion that include cell wall polysaccharides and polyphenols, lipids, and starch.

So, these substances were quite greatly getting doped or removed completely. However, such purification can often spearhead to sample loss. Hence, the use of self-packed Nano- Liquid chromatography columns for purification before matrix-assisted laser desorption or ionization Mass Spectrometry (MALDI-TOF MS) has been considered to work successfully.

3.5.1 Protein Extraction

The protein-pellet homogenization approach and the lysis buffer have a vital influence on protein solubilization and parting in classical proteomic research of plants. Most of the protocols being used in the extraction of proteins from plant tissues are being based on either phenol or tricarboxylic acid cycle /acetone precipitation methods. The tricarboxylic acid cycle/ acetone procedure has some limitations.

For instance, the resulting pellet may be quite difficult to dissolve nucleic acids higher than nearly 20 nucleotides and in most of the cases proteins can be hydrolyzed by tricarboxylic acid cycle.

Protein removal and sample preparation, particularly from recalcitrant tissues such as grape leaf, cucurbit pericarp, pine needle generally containing a high level of interfering substances, such as phenolic compounds, proteolytic and oxidative enzymes, terpenes, organic acids, and carbohydrates are among the most difficult aspects of proteomic analyses.

It has been projected in recalcitrant plant tissues and others that tricarboxylic acid cycle -acetone–phenol-based method is more effective than tricarboxylic acid cycle/acetone precipitation alone prior to protein solubilization.

However, various protein extraction and solubilization buffers with a diversity of chemical compositions and concentrations have been used, such as Mg/NP-40 buffer and Tris Buffer. The use of Phenylmethanesulfonylfluoride fluoride, a protein inhibitor, and many other protease inhibitors, function to minimize the proteolysis of proteins during isolation.

3.5.2 Protein Separation

Parting of a complex solution of proteins is being carried out by approaches which comprise of the pI, Mr, solubility, and the relative profusion. 2-DE, which comprises of Index of Economic Freedom from the first dimension and SDS-PAGE from the second dimension, allows this parting to happen. In addition to making sure the separation of a complex solution of proteins, 2-DE also consents for proteins that have done PTMs to be further isolated for structural analyses using a mass spectrometer.

The MALDI-TOF MS, MS alone, electrospray ionization (ESI) or Edman micro sequencing is used by the mass spectrometer to analyze the structures of the proteins.

In the same context, 2-DE MS consist of, sample preparation and protein solubilization, protein parting protein identification and quantification, computer-assisted research of 2-DE designs, protein detection and characterization, and two-dimensional protein database creation.

When researching protein spots on a 2-DE gel in-depth, the resolve is often getting restricted due to factors such as size, abundance hinder by denaturation and intermolecular interaction, and other electrophoretic properties.

Therefore, to improve the resolution and sensitivity, it has been figured helpful to divide the whole proteome into discrete sections, including organelles, subcellular compartments, and multi-protein complexes.

3.5.3 LC-MS-based Qualitative Analysis of Plant Systems

A completely sequenced genome basically aids in the procedure of obtaining outputs from the proteomic analysis. However, only a few eukaryotic organisms' genomes have been sequenced. There have been one thousand seven hundred and ninety-two eukaryotic genomes sequence dealing in comparison to the twenty-nine three hundred and sixty-nine sequenced prokaryotic genomes used for research about evolutionary processes, such as lateral gene transfer.

In comparison, another genomic study has its main aim on model organisms such as the human, rat, drosophila, *A. thaliana*, and rice. Recent research which has brought forth label-free researches of the plant proteome. However, a label-free method is not without problems, mainly when concerning plant samples which contain high abundance proteins and high amounts of coeluting peptides during LC-separation –like in cell lysates.

3.5.4 Biological Mass Spectrometry-Based Quantification

Biological mass spectrometry which is the full form used in place of MS is the use of a mass spectrometer in proteomics, which comprises of proteins and now peptides in this stage of the process that has been detached by SDS-PAGE.

These are then being meet up to the mass spectrometer via MALDI or ESI. After being introduced to the spectrometer, the mass-to-charge whose denotation is m/z of peptides are being quantified and been generating MS spectra.

Peptides are commonly disjointed by collision-induced dissociation, which wreckages the ions of peptides. MS-based analysis of proteins relies on two approaches: label-free and stable isotope labeling.

Label-free protein quantification is being branched into techniques being relying on either the intensity of the peptide ion in the mass spectrometer or the number of scans, the latter needs too much data to be efficient for Post-translational modification classification.

Stable isotope labeling is characterized by chemical labeling that rely on sulfhydryl groups of cysteine, free amines, or by enzymatic labeling performed on intact plant tissues or liquid cultures. Plants are often labeled with ^{15}N along with $K_{15}NO_3$ as the only nitrogen source during growth in soil or in cell suspension cultures enabling a large quantity of peptides enriched in ^{15}N to be identified, ninety-eight percent of which will be incorporated.

An alternative method called SILAC which is the acronym whose full form is Stable Isotope Labelling of Amino acids in cell culture is being used in which the cells are supplied with labeled amino acids, instead of ^{15}N, yielding eighty percent of the quantified proteins being incorporated, because plant cells already synthesize amino acids. Therefore, a stable isotope labeling technique using ^{2}H, ^{13}C, ^{15}N, or ^{18}O isotopes was developed for MS-based quantitative proteomics.

More than one dimension inducing protein identification technology which is the full form used in place of MudPIT uses a mixture of liquid chromatography which is either LC-MS/MS and high performance liquid chromatography (HPLC), improves the parting of peptide solution using strong cationic exchange and two dimensional liquid chromatography.

Thus, by refining sample parting methods along with efficient techniques of peptide mixture separation, the overall analysis of proteome can be enhanced to identify proteins of all functional and physical classes, which are found in low abundance.

3.6 PROTEOMIC EVALUATION OF GENETICALLY MODIFIED (GM) CROPS

Several types of research recommended the request of proteomics in identifying unintended effects in genetically significant modified harvests and variation in gene expression outlines of transgenic harvests. Transgenic linking may end up in deletions, insertions, and rearrangements, which in turn may have harmful effects.

These can be for cast by aimed analysis of Trans genetic linking sites and related metabolites; though they do not most necessarily forecast changes in food composition and quality, therefore, transcriptomics profiling is being limited in assessing unintended effects.

Proteomic technologies have also been applied to investigate gene silencing products in transgenic research. Substantial suppression of GlymBd 30 K, a dominant allergen of soybean seed, by a reverse genetic approach has been confirmed by proteomic analysis.

No chief variations were seen in the polypeptide pattern of the transgenic seeds were compared with the proteome map of the non-transgenic seeds. This might suggest that transgenic-induced gene-silencing technique quite perfectly removed the allergen from soybean. Several other allergen proteins comprise of profiling in olive pollen, which projected polymorphism_related to physiochemical cracks in answer to different stress or physiological conditions.

With the commercialization of GM harvests, these unintended effects are one of the most unconventional issues on the biological safety of GM harvests. The cumulative usage of "omics" technologies significantly provides to our already researched understanding of the biological safety of transgenic, since a systematic molecular analysis of GM crops is needed to address the unintended genetic effects.

3.7 MODIFICATION-SPECIFIC PROTEOMICS IN PLANT BIOLOGY

In the procedure of protein biosynthesis, post-translational modifications (PTMs) including phosphorylation, glycosylation, sulfation, prenylation, acetylation and ubiquitination are quite significant in changing protein structures, which comprise stability and localization, by binding to amino acids.

Researching and analyzing with Post-translational modifications projected that the functions of a protein are well conserved. However, the Post-translational modification bases may not be static over an evolutionary wide range time period.

Post-translational modification is not quite static under normal circumstances for proteomics or MS, thus making them quite difficult to research. However, they control many cellular functions, such as protein activity and signaling throughout the cell.

Post-translational modification is recognized by purification or protein separation using polyacrylamide gel electrophoresis (PAGE). The function of PTMs is currently being studied by analyzing the cell response to stress.

3.7.1 Phosphorylation

Reversible protein phosphorylation affects metabolism, the cell cycle, stress, hormonal responses, stomatal closure, development, as well as cytokinesis in the plant cell. Currently, plant phosphocreatine studies are carried out through Immobilized Metal Chromatography- IMAC, precipitation, metal oxides, such as TiO_2 and ZrO2.

Some proteins may be figured out in more than one organelle while others may be aimed to membranes with the help of lipid anchors. The research of subcellular fractionation makes it quite simpler to understand specific phosphorylation proceedings because the site can be determined. Phosphorylation has been shown to be involved in the cell's answer to fungal elicitors.

3.7.2 Ubiquitylation

Ubiquitylation is solely responsible for the classification of proteins that are required to plummet. It is a procedure where the C-terminal carboxyl group of a Ubiquitylation polypeptide is coagulated to the aimed protein with the help of a ε-amino group of one or several precise lysines remains within the target protein.

It requires an Adenosine Triphosphate -dependent Ubiquitylation activating enzyme which is also denoted by E1, a Ubiquitylation Conjugation enzyme which is also denoted E2, and a Ubiquitylation ligase which is also denoted E2, which recognizes substrates.

3.7.3 SUMOylation

SUMO which is the acronym for Small Ubiquitin-related Modifier, is as effective as a polySUMO link and is being involved in the transcriptional directive, Deoxyribonucleic Acid reparation, nuclear transport, mitochondrial fission, and regulation of receptors at the plasma membrane.

The enzymatic process involved in SUMOylation is similar to that used in ubiquitylation except "Small Ubiquitin-related Modifier required to be proteolytically processed by the Sentrin-specific Protease (SENPs) to expose the C-terminal GG before it can be activated".

In fact, in yeast and mammalian cell studies, there is increasing evidence that ubiquitylation and Small Ubiquitin-related Modifierylation pathways interact with each other.

Poly Small Ubiquitin-related Modifier chains are formed from SUMO small ubiquitin-like modifier links by proteins which comprise Small Ubiquitin-related Modifier interaction motif (SIM), i.e. the SUMO consensus modification motif (ΨKxE/D) in response to processes, such as Deoxyribonucleic Acid damage or meiosis.

Researched by Bruderer suggest that until now, more than three hundred putative polySUMO conjugates from cultured eukaryotic cells have been identified. Furthermore, both poly Small Ubiquitin-related Modifier and Small Ubiquitin-related Modifierylation modify proteins at specific stoichiometry.

3.8 APPROACHES AND CHALLENGES IN CROP PLANT PROTEOMICS

Along with the accomplishment of genome sequences in the model species such as monocotyledonous crop plant rice (*Oryza sativa* ssp. *japonica* and ssp. *indica*), dicotyledonous plant *Arabidopsis thaliana*, *Medicago truncatula* and *Lotus japonicus,* cereal crops *Zea mays,* and *Hordeum vulgare*, model grass *Brachypodium distachyon*, legume species soybean (*Glycine max*), and several other crop species, a focused attention has been given to linking genomic data as well as transcriptomic profiles to the temporal and spatial expression, biological and functional network of proteins.

In addition to that, dealing with dynamic and complex plant proteomes, it is vital to select appropriate proteomic approaches that target the recognition of proteins together with their modification that may eventually contribute to the improvement of crops.

In the past few years, quantitative proteomic studies applying high resolution as well as mass accuracy instruments have been majorly contributing to the understanding of growth in plants, their development and interactions with the environment.

Such kind of capability is very effective for crops as it contributes to increasing the nutritional value and production as well as to understand mechanisms of adaptations of crops to abiotic stresses responses.

REFERENCES

1. Agrawal, G., Sarkar, A., Righetti, P., Pedreschi, R., Carpentier, S., Wang, T., Barkla, B., Kohli, A., Ndimba, B., Bykova, N., Rampitsch, C., Zolla, L., Rafudeen, M., Cramer, R., Bindschedler, L., Tsakirpaloglou, N., Ndimba, R., Farrant, J., Renaut, J., Job, D., Kikuchi, S. and Rakwal, R. (2013). A decade of plant proteomics and mass spectrometry: Translation of technical advancements to food security and safety issues. Mass Spectrometry Reviews, [online] 32(5), pp.335-365. Available at: https://www.ncbi.nlm.nih.gov/pubmed/23315723 [Accessed 10 Jun. 2019].
2. Buts, K., Michielssens, S., Hertog, M., Hayakawa, E., Cordewener, J., America, A., Nicolai, B. and Carpentier, S. (2014). Improving the identification rate of data independent label-free quantitative proteomics experiments on non-model crops: A case study on apple fruit. Journal of Proteomics, [online] 105, pp.31-45. Available at: https://www.ncbi.nlm.nih.gov/pubmed/24565695 [Accessed 10 Jun. 2019].
3. Bykova, N., Hoehn, B., Rampitsch, C., Banks, T., Stebbing, J., Fan, T. and Knox, R. (2011). Redox-sensitive proteome and antioxidant strategies in wheat seed dormancy control. PROTEOMICS, [online] 11(5), pp.865-882. Available at: https://www.ncbi.nlm.nih.gov/pubmed/21280218 [Accessed 10 Jun. 2019].
4. Chakraborty, S. and Newton, A. (2011). Climate change, plant diseases and food security: an overview. Plant Pathology, [online] 60(1), pp.2-14. Available at: https://www.natureasia.com/en/nindia/article/10.1038/nindia.2015.121.
5. Eldakak, M., Milad, S., Nawar, A. and Rohila, J. (2013). Proteomics: a biotechnology tool for crop improvement. Frontiers in Plant Science, [online] 4. Available at: https://www.ncbi.nlm.nih.gov/pmc/articles/PMC3584254/ [Accessed 10 Jun. 2019].
6. Gillet, L., Navarro, P., Tate, S., Röst, H., Selevsek, N., Reiter, L., Bonner, R. and Aebersold, R. (2012). Targeted Data Extraction of the MS/MS Spectra Generated by Data-independent Acquisition: A New Concept for Consistent and Accurate Proteome Analysis. Molecular & Cellular Proteomics, [online] 11(6), pp. O111.016717. Available at: https://www.ncbi.nlm.nih.gov/pubmed/22261725 [Accessed 10 Jun. 2019].
7. Hu, J., Rampitsch, C. and Bykova, N. (2015). Advances in plant

proteomics toward improvement of crop productivity and stress resistancex. Frontiers in Plant Science, [online] 6. Available at: https://www.frontiersin.org/articles/10.3389/fpls.2015.00209/full [Accessed 10 Jun. 2019].

8. Jacoby, R., Millar, A. and Taylor, N. (2010). Wheat Mitochondrial Proteomes Provide New Links between Antioxidant Defense and Plant Salinity Tolerance. Journal of Proteome Research, [online] 9(12), pp.6595-6604. Available at: https://www.ncbi.nlm.nih.gov/pubmed/21043471 [Accessed 10 Jun. 2019].
9. Khoranhlai, K. (2015). Advances in Proteomics and Bioinformatics in Agriculture Research and Crop Improvement. Journal of Proteomics & Bioinformatics, [online] 08(03). Available at: https://www.omicsonline.org/open-access/proteomics-and-bioinformatics-in-agriculture-research-and-crop-improvement-jpb.1000351.php?aid=40357 [Accessed 10 Jun. 2019].
10. Pascovici, D., Gardiner, D., Song, X., Breen, E., Solomon, P., Keighley, T. and Molloy, M. (2013). Coverage and Consistency: Bioinformatics Aspects of the Analysis of Multirun iTRAQ Experiments with Wheat Leaves. Journal of Proteome Research, [online] 12(11), pp.4870-4881. Available at: https://www.researchgate.net/publication/256477861_Coverage_and_Consistency_Bioinformatics_Aspects_of_the_Analysis_of_Multirun_iTRAQ_Experiments_with_Wheat_Leaves [Accessed 10 Jun. 2019].
11. Picotti, P., Bodenmiller, B. and Aebersold, R. (2012). Proteomics meets the scientific method. Nature Methods, [online] 10(1), pp.24-27. Available at: https://www.ncbi.nlm.nih.gov/pubmed/23269373 [Accessed 10 Jun. 2019].
12. Smith, R., Mathis, A., Ventura, D. and Prince, J. (2014). Proteomics, lipidomics, metabolomics: a mass spectrometry tutorial from a computer scientist's point of view. BMC Bioinformatics, [online] 15(S7). Available at: https://bmcbioinformatics.biomedcentral.com/articles/10.1186/1471-2105-15-S7-S9 [Accessed 10 Jun. 2019].
13. Uvackova, L., Skultety, L., Bekesova, S., McClain, S. and Hajduch, M. (2013). MSE Based Multiplex Protein Analysis Quantified Important Allergenic Proteins and Detected Relevant Peptides Carrying Known Epitopes in Wheat Grain Extracts. Journal of Proteome Research, [online] 12(11), pp.4862-4869. Available at: https://pubs.acs.org/doi/

abs/10.1021/pr400336f [Accessed 10 Jun. 2019].

14. Uvackova, L., Skultety, L., Bekesova, S., McClain, S. and Hajduch, M. (2013). The MSE-proteomic analysis of gliadins and glutenins in wheat grain identifies and quantifies proteins associated with celiac disease and baker's asthma. Journal of Proteomics, [online] 93, pp.65-73. Available at: https://www.ncbi.nlm.nih.gov/pubmed/23268118 [Accessed 10 Jun. 2019].

15. Wu, D., Shen, Q., Qiu, L., Han, Y., Ye, L., Jabeen, Z., Shu, Q. and Zhang, G. (2014). Identification of proteins associated with ion homeostasis and salt tolerance in barley. PROTEOMICS, [online] 14(11), pp.1381-1392. Available at: https://onlinelibrary.wiley.com/doi/pdf/10.1002/pmic.201300221 [Accessed 10 Jun. 2019].

4

Role of Proteomics and Animal Food Source

CONTENTS

The chapter of role of proteomics and animal food source talks about the study of the proteomics as well as the various types of applications of the proteomic tools in the life sciences. This chapter also mentions the several number of applications of proteomic tools which are used to measure the quality of the food and safety of the food products. This chapter explains several types of proteomics technologies such as 2D PAGE MS, Mud PIT, top-down Approach, and SELDI. This chapter also provide highlights on the application of proteomics in the food technology and food biotechnology such as proteomics as a tool for product and process validation and optimization. This chapter also addresses technical aspects of the animal proteomics which includes sample selection and preparation, the recent trend to the mass spectrometry, gel-based proteomics and bioinformatics in animal proteomics. This chapter talks about the plants and animals as a source of food along with the important nutrients for the vegetarians such as calcium, iron, proteins, vitamin B12, and vitamin D. This chapter also explains the essential characteristics of food by ensuring the quality of the life. This chapter describes that some of the nutrients are more abundant in the animal protein source.

4.1 INTRODUCTION

Proteomics is the study of comprehensive protein profile of a cell or a tissue or the fluid which is present in the body at a given physiological or pathological condition. The proteome of a cell can be well-thought-out as the connection between the genome and the function of a cell. Proteins are very important factor or element of a complicated biochemical and metabolic set of pathways that explains any kind of organism and the biology of protein.

Proteins perform a very crucial role in a different range of metabolic factors in all living organisms, from structural roles to the energy metabolism regulation, from parasite tolerance to the disease response. Proteins are the most important element of every productive standard of farm animals, which lies from feed quality to the product certification, from the creation of vaccines to the studies of the immune response, dairy product processing.

The main objective of the Proteomics is to analyze several proteins at a time dissimilar from the conventional identification of one protein at a time. In contrast to "genome" there is no single, stationary proteome in any organism as an alternative there are several dynamic sets of proteins in various cells and tissues that show fluctuation as an outcome to several numbers of situations for example stress or infectious processes.

From now, the categorization of the entire proteome of a multicellular organism is a very difficult job because of the extended dynamic range of protein expression. The most striking aspect of the farm animal proteomics is the variety in the farm animals, tissues, organs and fluids which are present in the body along with the products and their processing settings.

This complication is increased if the farm animal proteomics consist of a cumulative number of diseases and several types of agents are causing the disease, for example, bacteria, viruses and parasites along with the possible vectors of these types of diseases namely ticks, flies and mosquitoes.

The main aim of increasing livestock is to make food or protein to fulfill the human hunger requirements. The nutritive value and edibility of the products which are obtained from animals are determined with the help of the kind of proteins which are present in it.

A very crucial application of the proteomics with respect to the food produced from the livestock is in the modification or changes of the proteins following harvest such as the changes in the muscle proteins as it becomes meat and the modification of milk proteins through the course of the process of the formation of cheese.

There are mainly two types of methods for the categorization of the proteome which are known as comparative proteomics and mapping proteomics. Mapping proteomics which has the target to characterize and make wide-ranging databases or cellular proteomes. This is a very huge task, partly because of the complicated diversity of modification forms of most proteins possess and constant alterations in the proteome with time.

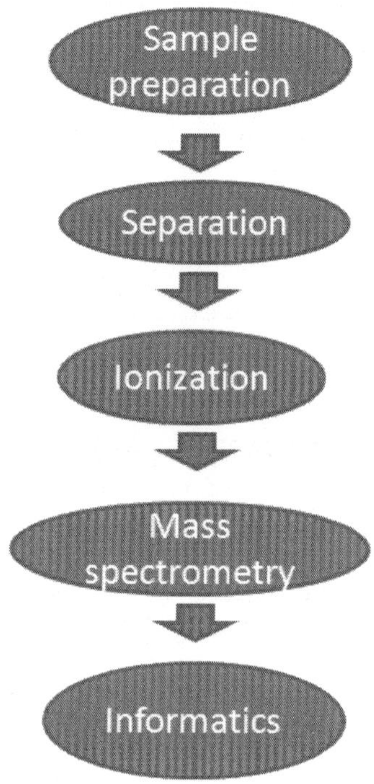

Figure 4.1: A flowchart of the proteomics.

Source: https://upload.wikimedia.org/wikipedia/commons/f/f3/Flowchart_of_proteomics.gif

Relative proteomics has the target to describe the biological mechanisms that form the connection between the noticeable phenotypes and genotypes, in that way making a moment with the help of the moment snapshots of cellular responses at the protein level.

Proteome analysis depends upon the four most important elements which are protein separation, identification, characterization and quantification essentially resorting to mass spectrometry. The proteome experimental methods have developed from the traditional identification of all proteins to include functional proteomics (quantification and identification of differentially expressed proteins among distinct conditions), identification and characterization of protein-protein interactions and characterization of protein post-translational modifications.

Evaluation of the cellular proteome is a very difficult task because of the complication of proteins in tissues and cell lysates. There are two comprehensive groups of proteomic methods in proteomic research, gel-based assays and shotgun proteomics which are used in the present interval of time.

4.2 APPLICATION OF PROTEOMIC TOOLS IN FOOD QUALITY AND SAFETY

With the modification in time as well as the development in the technology, increase in the population has elevated the need of the food, which stresses on the growth of advanced food production and processing techniques. Expanding to the requirement of the food science research which is concentrating on the improvement of the nutritional quality of food as well as the safety.

The new innovative methods are known to classify the quality and safety aspects of the food. One of these types of tactics is known as '–omics'-based research, which is increasing its own significance for the acknowledgment of the basic procedures in the food science, which is known as Food omics, stated by the researcher named as Cifuentes, in the year of 2009.

Food omics is defined as the studies on food with the help of the application of improved omics tactics. Foodomics consist of genomics, transcriptomics, epigenetics, proteomics, peptidomes, and or metabolomics to evaluate the safety of the food, quality of the food, traceability of the food and finding new bioactive elements in the food.

The word "Proteomics" was coined with the help of the Marc Wilkins at Siena conference in the year of 1994 and the term proteome was shortened from the "Protein complement of the Genome" meaning the entire set of the proteins expressed with the help of the cell, stated by the Wilkins, in the year of 1996. Proteomics could be explained as "the large-scale analysis of proteins", according to the two researchers named as Pandey and Mann, in the year of 2000.

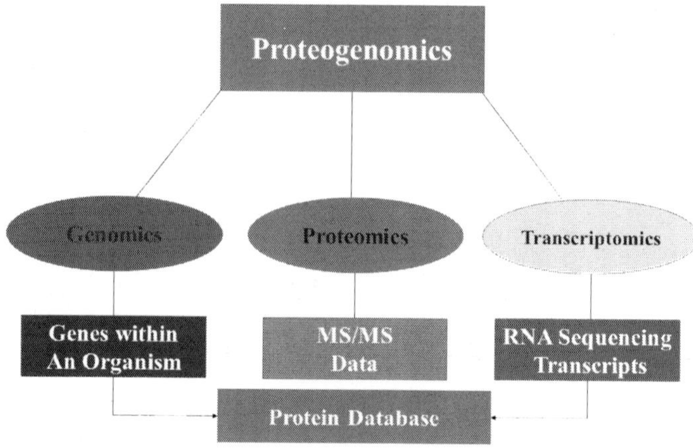

Figure 4.2: Applications of the Proteomic tool in the food quality and safety of food.

Source: https://upload.wikimedia.org/wikipedia/commons/3/3a/Proteogenomics_Image_2.png

"Proteomics includes not only the identification and quantification of proteins, but also the determination of their localization, modifications, interactions, activities, and, ultimately, their function", stated by the Fields, in the year of 2001.

The workflow of proteomics methodology

- Sample collection, handling and storage.
- Protein separation (Ingel or OFFGEL).
- Protein identification (peptide mass fingerprinting)
- Protein characterization (amino acid sequencing).
- Bioinformatics (cross-reference of protein).

Proteomic methods are categorized into six sets which are expression proteomics, protein-protein interactions, functional proteomics, structural proteomics, proteome mining, and post-translational modifications, according to the researcher named as Carbonaro, in the year of 2004.

Proteomics is a new promising method to classify protein which is present in the food matrix and to examine protein–protein interactions in both raw as well as processed foods, along with the interactions between proteins and other food components, according to the researcher named as Kvasnicka, in the year of 2003.

Proteomics provides delicate information on the alteration in the structure of the protein occurring at specific amino acid residues through the course of processing events and supports to anticipate the quality and stability of food product. In proteomics, evaluation of the proteins and peptides is the most important element and the most difficult task with respect to the separation of them from the difficulty and dynamic concentration range in food.

The prime use of the proteomic method is in studies that involve the alteration in the quality of the food stated by the Werf, et al. (2001), in the year of 2001, distinguishing the bioactive compounds, recognizing the ingredients in the functional foods and nutraceuticals, according to the Galvani et al. (2001), 2001, classification of the proteins present in the rice, according to Koller et al., (2002), , molecular marker for physiology and pathobiology, according to Bouley et al. (2005), integrated food omics, according to the Ibá˜nez et al. (2013), 2013 , in-vivo protein digestion trail, stated by the Carbonaro (2004) 2003, identification of allergic proteins, according to Beyer et al. (2002) , diagnosis, targeted treatment and vaccine/ drug development.

4.3 PROTEOMICS TECHNOLOGIES

Two-dimensional polyacrylamide gel electrophoresis (2D-PAGE) and several type of chromatography methods have long been used as useful methods for the separation of protein.

Nevertheless, just with the development of delicate ionization strategies, for example, grid helped laser desorption ionization (MALDI) and electrospray ionization (ESI), in the late timespan of year the 1980s to vaporize peptides and proteins, has their assessment of proteins haves become conceivable with the assistance of MS.

The partition of ionized peptides or proteins in the high vacuum of the MS depends on the modifications in their mass-to-charge proportions, with time-of-flight (TOF) and electric particle trap as significant mass-particular analyzers in MALDI and ESI instruments, separately.

Over the timeframe of recent years, MS has contributed an exceptionally urgent job in the headway or improvement of the proteomics abilities because of its rapid, affectability, particularity, mass goals capacity, and mass exactness.

The prevalent intensity of MS in the proteome examination can be additionally improved when MS (as an identifier) is joined with LC-GC or when MS-MS is utilized.

There are as of now 4 noteworthy proteomics advancements that include either explicit processing of proteins [2D-PAGE MS and multidimensional protein ID innovation (MudPIT)] or direct assessment of proteins after their chromatographic detachment [the top-down methodology and surface-improved laser desorption ionization (SELDI)].

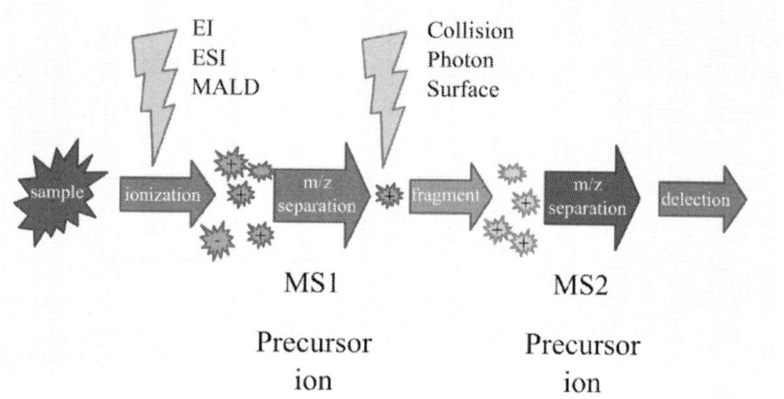

Figure 4.3: Technologies which are used in the field of proteomics.

Source: https://upload.wikimedia.org/wikipedia/commons/thumb/6/68/MS_MS.svg/1280px-MS_MS.svg.png

The choice of technology depends on the goal of the study and facility availability. However, all methods require sample preparation, protein separation, MS analysis, and protein identification. The last mentioned is based on peptide mass fingerprints or spectra of fragmented ions of intact proteins and a search of the published databases.

4.3.1 2D-PAGE MS

This is the traditional and most commonly applied proteomics technology. Proteins are separated in the first dimension connecting to the isoelectric focusing, and in the second dimension consisting of SDS-PAGE, depending on their isoelectric points (net charges) and sizes (molecular weights), respectively.

The creation of the immobilized pH gradients has permitted the narrow pH intervals (for example, 4–7 vs. the standard 3–10) and enhanced the dependability and reproducibility of the protein separation with the help of the isoelectric focusing. After the 2D-PAGE separation, proteins are stained with the colloidal Coomassie blue, silver, or fluorescent dyes.

In the recent interval of time, fluorescence-based variance gel electrophoresis (DIGE) was created to remove the gel-to-gel variation, which is consisting of the addition of fluors 1 and 2 to the control and treatment samples, respectively.

Computerized quantitative image evaluation is performed with the help of the special software programs made specially to perform the task. In-gel digestion is done with the help of the enzymes (usually with trypsin) or chemically (with the help of the cyanogen bromide), followed by the MALDI-TOF MS evaluation of the generated peptides.

2D- PAGE approach is considered as the most effective technology for the separation of the protein, as it can be used to carry out relative side by side quantification of proteins on gel images, and recognition of protein isoforms and post-translational modifications (for example phosphorylation, hydroxylation, methylation, glycosylation, acetylation, and oxidation).

The disadvantages of 2D- PAGE are challenged efficacy in resolution (for example, membrane proteins and some plasma proteins) variations in isoelectric points, molecular weights, abundance, and hydrophobicity, and challenges with automation or complete quantitative information.

4.3.2 MudPIT

This type of approach is consisting of initial digestion of proteins with the help of the specific protease (commonly trypsin). The produced peptides are separated with the help of the strong cation exchange and reversed-phase HPLC, followed with the help of the MS evaluation. Because the ESI ionization source is compatible with a liquid sample, ESI-MS/MS is every often suitably combined with HPLC.

As well as with MudPIT, isotope labeling (as well as $^{15}N/^{14}N$, $^{18}O/^{16}O$, or isotope-coded affinity tags) can be applied to differentially label proteomes in control and treated samples, which produces quantitative proteomics information. Another quantitative technique (which is known as the Absolute Quantitative Analysis) consist of the addition of a ^{13}C-labeled peptide to a protein digestion mixture for determining peptide recovery through the course of the sample processing.

MudPIT not only incapacitates the shortcomings of 2D-PAGE MS but also offers the following profits such as removal of the time-consuming step for protein separation; high sensitivity and necessity of small sample size; and versatile mechanisms for peptide separation. A very crucial or main drawback of MudPIT is its incapability to offer information regarding the protein isoforms or post-translational modifications.

4.3.3 Top-down Approach

This involves the utilization of an acid-labile detergent in gel electrophoresis to resolve proteins first followed by reversed-phase HPLC and MS (sometimes known as ESI-MS/MS). In the recent interval of time, Fourier transform ion cyclotron resonance has been employed to separate protein ions with various masses, but the same mass-to-charge ratios exhibit the same cyclotron frequency.

The primary usefulness of this approach is the range, added data on isoforms or post-translational modifications as well as analyzing membrane proteins that are soluble in nature.

This is more specifically crucial when the genes of interest have not been sequenced. Even though the top-down method is quickly increasing its recognition, it is at present not supported with the help of the software packages for the processing of complicated data sets and may not have the potential to sequence all amino acid residues in proteins containing the cyclic heme.

4.3.4 SELDI

This uses a Protein Chip Array based on ion-exchange or LC and antibody- or substrate-based affinity directly from the original source material. This chip allows for resolving proteins from the complicated protein mixtures.

The captured proteins are evaluated with the help of the laser desorption or ionization which is based on MS-TOF. The prime benefits of the technology SELDI are simple preparation of samples, lowered sample complication, rapidity and suitability for low abundance proteins (for example transcription factors and most cellular proteins).

Application of Proteomics in Food Technology and Food

Biotechnology:

Application of Proteomics in Food Technology and Food Biotechnology:

4.4 APPLICATION OF PROTEOMICS IN FOOD TECHNOLOGY AND FOOD BIOTECHNOLOGY

The utilization of proteomics for the procedure development and approval in food innovation and biotechnology just as steady quality control of the beginning materials and last items was toward the starting moderately restricted first. There were just a few presentations in the segments 'Biotechnology points of view's and 'Proteomics in biotechnology' at HUPO World Congress five years prior, some of them are truly adapting to the utilization of the proteomics. In the time interim of a years ago, it has adjusted in all respects rapidly so proteomics innovation is routinely connected, and the terms 'modern procedure proteomics' and 'mechanical proteomics' are presently normally utilized

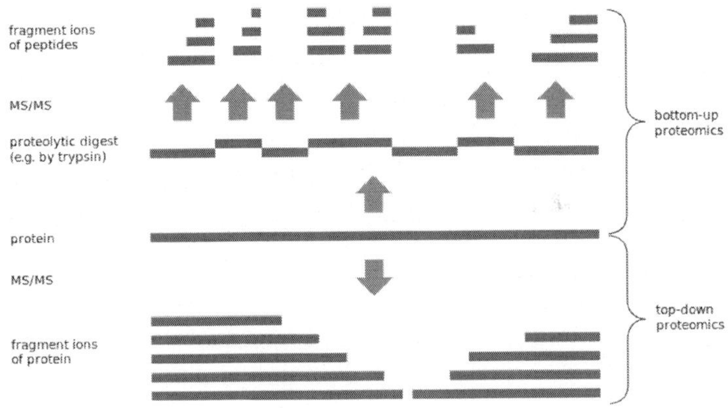

Figure 4.4: Application of proteomics in food technology and food biotechnology.

Source: https://upload.wikimedia.org/wikipedia/commons/thumb/f/f0/Bottom-up_vs_top_down.svg/800px-Bottom-up_vs_top_down.svg.png

4.4.1 Proteomics as a Tool for Product and Process Validation and Optimization

In a spearheading work, played out a deliberate proteomic assessment just as a plant-scale wet fractionation process, just as a plant-scale wet fractionation process, was done of alfalfa biomass. It was likewise clarified that through the course of the biomass handling a specific degree of cell regulation is as yet safeguarded, for instance, enlistment of heat and redox stress proteins. Proteolytic digestion of proteins and different variances in meat likewise start through the course of the capacity and the main handling venture of protein-rich materials, for instance, porcine meat.

Improvements in the conventions for food handling have prompted an abatement in the assembling time and enhancement of item quality. The increment in the limit of the production likewise boosts the prerequisite for better procedure control. Programming driven PC control frameworks, for instance in the milk or meat handling industry, have made it a lot simpler just as quicker to adjust the parameters through the course of the preparing and generation cycles

Proteins are mainly accountable for the physical appearance of several numbers of food products through the course of the manufacturing process. Physicochemical properties, for example, viscosity, thermal conductivity and vapor pressure, and also nutritional and sensory properties of milk, meat and cereal-derived products rely on their protein configuration and content.

In the products which are derived with the help of the wheat flour, the optimal physical appearance is evaluated with the help of the gluten proteins; in milk and products made of milk, the dominating protein is casein. The proteomics-based method for validation of a process for the production of wheat-based foods both proteins, as well as protein groups, are well categorized.

Protein compositions of other foods like meat and meat products, or fruit and vegetables are more complex with the properties dependingon one or more than one highly rich protein.

A noteworthy quantity of pork and beef is the importance of juiciness and texture as vital organoleptic physical appearance. Proteomics shows that these factors are dependant on structural proteins for example myosin, actin, desman and tubulin.

In a semi-quantitative comparison, which is based on the evaluation of concentration of several types of protein or peptide spots in two-dimensional

2D electrophoresis, a researcher named as Laville, has identified fourteen various types of proteins that are a kind of 'candidate biomarker' for shear force values of cooked meat.

Additional or higher studies about the meat texture and drip loss were also made. A renowned researcher named as Sayd holds the enzymes of oxidative metabolism as the force behind meat quality by color enhancement in the sarcoplasmic reticulum of pig muscle.

Muscle mitochondria are also extremely delicate to protein carbonylation. With the help of application of a complicated labeling strategy, more than two hundred carbonylated proteins were spotted. Other oxidative changes, for example, nitrosylation and hydroxylation were also spotted in several numbers of carbonylated proteins.

This discovery offers additional evidence of the susceptibility of muscle mitochondrial proteins to oxidative damage. The properties are dependant on treatment and storage during production as seen in fish muscle proteins,.

4.5 PROTEOMICS IN VETERINARY MEDICINE AND ANIMAL SCIENCE

Proteomic researches with respect to the veterinary medicine and animal health have elevated in recent time but is still a minor element of the large body of reports in the proteomics canon. Investigation of the proteomes of tissue and biological fluids from the animals can be for health and diseases in veterinary medicine, in which case they are parallel with the investigation of human disease.

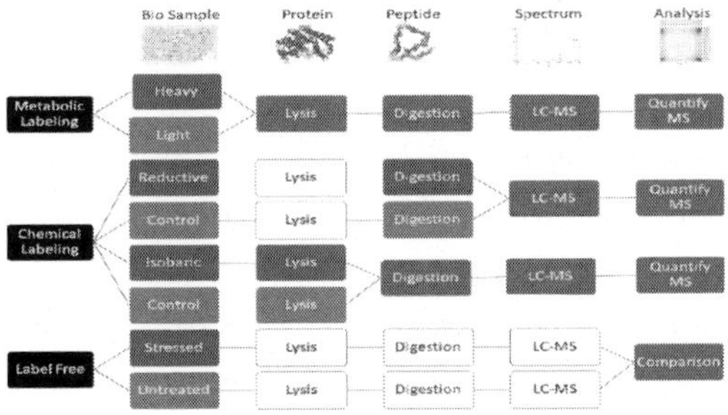

Figure 4.5: Mass Spectrometry Quantitative Proteomic Workflows.

Source: https://upload.wikimedia.org/wikipedia/commons/2/23/Mass_Spectrometry_Quantitative_Proteomic_Workflows.png

Nevertheless, there are some extra aspects of comparative proteomics that provide extra or additional scientific value to the proteomics in species for example cattle, dogs, poultry, cats, pigs, horses, and fish and they are:

-Animal proteomics is an area of examination with the use to the biology and pathology of domestic organisms offering appreciated understanding in central factors of each species.

-Relative proteomics permits variances as well as resemblances between the proteome in health and disease to be associated between species, producing attractive insight into how species proteomes have developed themselves.

-Experimental proteomics in the field of domestic animals has an essential benefit over the application of the rodents, for example with greater application of the several sampling in time-series studies and with non-invasive (such as milk, saliva, urine) or minimally invasive (such as serum, plasma) samples being available in the satisfactory volume for several numbers of analyses.

-Both experimental, as well as natural disease processes, can be examined with the help of the proteomics in the same species, different disease examines in the body of humans that can only be undertaken in patient samples from natural disease.

-Species other than rodents are more suitable as the models for human physiology , for instance, pig or dog is essential for the drug safety testing by innovative pharmaceuticals approach regulatory approval.

-Interaction between the proteomic phenotype and genetics in domestic animals can be supported with the help of the knowledge of the population genetics of species where periods of recorded breeding are an invaluable resource on proteome–genome interactions.

In the recent interval of time, there have been various excellent reviews with respect to the animal proteomics, which are dedicated to developing in the area over the last ten years and has been the subject of research . More willingly than repeat the information contained therein, this point of view is to inspire the greater contribution in this neglected niche of proteomics, where fascinating questions and hypothesis-driven investigations can be generated.

4.6 PLANTS AS A SOURCE OF FOOD

These are the major sources of foods with fruits and vegetables as well as crops, bushes, leaves and sometimes even roots.

Crops are vital seeds that are harvested for various food item sconsumed commonly in the diet such as bread or flour.

Animals such as cows or giraffes are herbivores while humans who only consume foods from plants are vegetarians. Let us look at examples of some foods we get from plants and their different parts.

- -Fruits: Bananas, Apples, Mangoes Oranges,
- -Leaves: Lettuce Spinach, Kale, Coriander,
- -Root: Turnips, Potato, Carrots,
- -Seeds: maize, wheat, Rice, nuts (peanuts, almonds)

4.7 ANIMALS AS A SOURCE OF FOOD

Other sources of food include animals such as cattle while a few animals are hunted) like deer or rabbits or even kangaroos (especially in Australia).

Some examples of such meat and their sources are:

- Mutton which is from lambs
- -Chevon which is goat meat
- -Chicken (poultry)
- -Bacon/pork from pigs
- -Beef from cows and buffalos
- -Venison from deer
- -Seafood includes fishes and other sea creatures like crabs, lobsters, oysters, etc. that are caught and consumed.

This is followed by dairy products such as milk, curd, butter and cheese . This include Bees yield honey: another source of food.

4.8 FOOD SOURCES OF 5 IMPORTANT NUTRIENTS FOR VEGETARIANS

Following the MyPlate tips can allow monitoring the nutritional requirements

4.8.1 Calcium

Vegetarians should consume a variety of calcium sources (with dairy products serving as rich sources) in order to meet daily requirements.

- -Low-fat or fat-free milk, yogurt and cheese
- -Fortified plant-based milk such as soy or almond
- -Fortified ready-to-eat cereals
- -Calcium-fortified juice
- -Calcium-set tofu
- -Some leafy green vegetables including collard greens, turnip greens and kale
- -Broccoli
- -Beans including soybeans, chickpeas and black beans
- -Almonds and almond butter

4.8.2 Iron

The consumption of a good source of vitamin C (citrus fruits, peppers, tomatoes) at each meal increases iron absorption.

Sources include:

- -Fortified breakfast cereals
- -Soybeans
- -Some dark leafy greens including spinach and chard
- -Beans
- -Eggs

4.8.3 Protein

This is found in both types of food: plant and animals.

Vegetarian sources of protein include:

- -Legumes such as beans, peas and lentils
- -Whole grains

- -Soy products
- -Nuts and nut butter
- -Dairy products
- -Eggs

4.8.4 Vitamin B_{12}

B_{12} is sourced from animal foods such as eggs and dairy products. An important point is for vegans who require fortification by a vitamin B_{12} (cobalamin) supplement.

Vegetarian sources of B_{12} include:

- -Vitamin B_{12}-fortified foods including nutritional yeast, soymilk, meat analogs, and ready-to-eat cereals (be sure to check the label).
- -Dairy products
- -Eggs

4.8.5 Vitamin D

Few foods are naturally high in vitamin D, but many dairy products are fortified with vitamin D in the United States. People who choose not to eat dairy products and who do not receive exposure to sunlight on a regular basis should discuss the need to take a vitamin D supplement with their health care provider.

Vegetarian sources of vitamin D include:

- -Eggs
- -Vitamin D-fortified soymilk, cow's milk, orange juice and ready-to-eat cereals

4.9 ESSENTIAL CHARACTERISTICS OF FOODS: ENSURING QUALITY OF LIFE

Competence of foods to withstand the human survival and well-being consists of not the only satisfactory amount, but also the nutrition , useful quality and safety of foods. Foods which is used up may encompass several types of nutrients, non-nutrients, anti-nutrients, and other functional constituents as well, for example, phytochemicals.

The nutritional competence of each population is directed with the help of the Dietary Reference Intakes, DRI (a new concept introduced with the help of the National Academy of Science, US, in the past known as Recommended Dietary Allowances, RDA), developed for the macro- and micronutrients. Necessities for nutrients have been recognized, methodically proven to have important meanings or purposes in the body of the humans and reviewed periodically as new evidence emerges.

The nutrient references are made with respect to the amounts of nutrients which is required daily with the help of the several age groups, sexes, and physiological groups (namely, pregnancy and lactation), to preserve the appropriate growth, functions, as well as health.

Foods are sources of nutrients such as carbohydrates, protein, fat, vitamins, and minerals as well. Energy is resultant which is derived largely with the help of the carbohydrates and fat, and from protein when carbohydrate and fat intakes are insufficient in amount. Most important food sources for the carbohydrates are known as cereal grains and starchy tubers and roots.

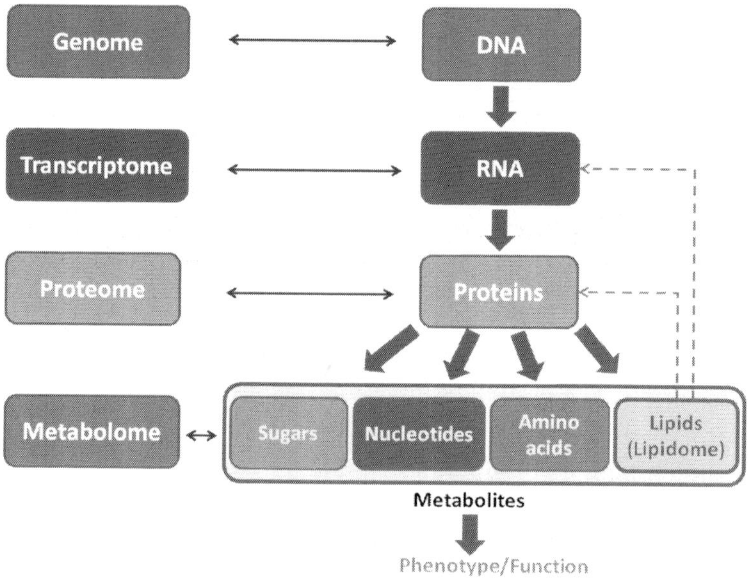

Figure 4.6: Essential characteristics of foods: ensuring the quality of life.

Source: https://upload.wikimedia.org/wikipedia/en/9/98/Metabolomics_schema.png

Animals, fish, and dried seeds (pulses and legumes) offers both proteins as well as fat. Vitamins and minerals are obtained from indingvegetables and fruits, nevertheless cereals, animal sources, and dried seeds also offer considerable quantities. Some types of tubers, for example, sweet potatoes, and their leaves also contain carotenoids.

Taro leaves (Colocasia esculenta), which is largely used in the regions of Asia and the Pacific, and cassava leaves (Manihot esculenta), which is used in the regions of humid tropical Africa, are very good sources of pro-vitamin A. Eggs from a several numbers of species, for example chicken, duck, quail (Asia) and guinea fowl (Africa), can offer comprehensive vital amount of amino acids.

They are also good sources of vitamins and minerals. Fish, nevertheless seemingly common, but it is still costly for some populations, only if available with the help of gathering from natural sources or aquaculture.

Nutrient and non-nutrient food constituents do not exist and function in isolation. They can synergistically interact to elevate the absorption, availability, and utilization, or antagonistically inhibit the availability of each other.

These relations are known as nutrient bioavailability. Consequently, to make nutritionally satisfactory meals, both the content and bioavailability of nutrients must be well-thought-out. Several nutrients and non-nutrients coexist in foods and foods are likely to be used up in combination.

Consequently, the interaction of nutrients, as well as non-nutrients from the same dishes or between the dishes eaten together, can be expected. For instance, high oxalate in foods secretes a negative effect on the absorption of calcium along with the iron. Calcium in spinach which coexists with oxalate is not willingly available when the oxalate to calcium ratio is larger than the value of two .

Foods that contain no utilizable calcium, but have excessive oxalate, can combine with the calcium from other foods eaten at the same time, according to the two researchers named as Noonan and Savage, in the year of 1999. Absorption of zinc can be combined with the help of the iron if the ratio of iron to zinc is larger than four .

Another crucial function of food is to offer physiologically functional ingredients, which avoids certain types of diseases and encourage good health. They are supposed to be effective in the stoppage and, or treatment of any type of disease for the betterment of health.

Physiological functions commonly consist of:

- bio defensiveness,
- the rhythm of physical condition,
- prevention of degenerative conditions due to aging, and
- prevention of certain diet-related diseases.

4.10 SOME NUTRIENTS ARE MORE ABUNDANT IN ANIMAL PROTEIN SOURCES

Of course, proteins are rarely found in isolation. They usually come with a wide variety of other nutrients.

Foods that contain animal protein tend to be high in several nutrients that are often lacking in plant foods.

These include:

4.10.1 Vitamin B12

Vitamin B12 is mainly found in fish, meat, poultry and dairy products. Many people who avoid animal foods are deficient.

4.10.2 Vitamin D

Vitamin D is found in oily fish, eggs and dairy. Some plants contain it, but the type found in animal foods is better used by your body.

4.10.3 DHA

Docosahexaenoic acid (DHA) is an essential omega-3 fat found in fatty fish. It›s important for brain health and is hard to get from plant sources.

4.10.4 Heme-iron

Heme-iron is predominantly found in meat, especially red meat. It is much better absorbed in the body than non-heme iron from plant foods.

4.10.5 Zinc

Zinc is mainly found in animal protein sources, such as beef, pork and lamb. It is also more easily absorbed and obtained from animal protein sources.

Of course, there are also plenty of nutrients found in plants that are lacking in animal foods. Therefore, eating balanced amounts of both is the best way to get all the nutrients you need.

REFERENCES

1. Almeida, A., Bassols, A., Bendixen, E., Bhide, M., Ceciliani, F., Cristobal, S., Eckersall, P., Hollung, K., Lisacek, F., Mazzucchelli, G., McLaughlin, M., Miller, I., Nally, J., Plowman, J., Renaut, J., Rodrigues, P., Roncada, P., Staric, J. and Turk, R. (2014). Animal board invited review: advances in proteomics for animal and food sciences. *Animal*, [online] 9(1), pp.1-17. Available at: https://www.ncbi.nlm.nih.gov/pmc/articles/PMC4301196/ [Accessed 10 Jun. 2019].
2. Jayasri, K., Padmaja, K. and Eswar Prasad, P. (2019). *Proteomics in Animal Health and Production*. [ebook] Tirupati, India: IOSR Journal of Agriculture and Veterinary Science. Available at: https://pdfs.semanticscholar.org/c692/7ed743e1e394149f1d6b7faf960fc42c446e.pdf [Accessed 10 Jun. 2019].
3. Ortea, I., O'Connor, G. and Maquet, A. (2016). Review on proteomics for food authentication. *Journal of Proteomics*, [online] 147, pp.212-225. Available at: https://www.sciencedirect.com/science/article/pii/S1874391916302792 [Accessed 10 Jun. 2019].
4. Shankaregowdanakoppalu Jagadeesh, D., Kannegundla, U. and Krishna Reddy, R. (2019). *Application of Proteomic Tools in Food Quality and Safety*. 5th ed. [ebook] Advances in Animal and Veterinary Sciences. Available at: http://nexusacademicpublishers.com/uploads/files/AAVS_5_5_213-225.pdf [Accessed 10 Jun. 2019].
5. Toldrá, F. and Nollet, L. (2013). *Proteomics in foods*. New York: Springer.
6. Toppr (2019). *Sources of Food: Plants and Animals, Concepts with Solved Questions*. [online] Toppr guides. Available at: https://www.toppr.com/guides/science/food-where-does-it-comes-from/sources-of-food/ [Accessed 11 Jun. 2019].
7. Wang, J., Li, D., Dangott, L. and Wu, G. (2006). Proteomics and Its Role in Nutrition Research1,2. *The Journal of Nutrition*, [online] 136(7), pp.1759-1762. Available at: https://academic.oup.com/jn/article/136/7/1759/4664604 [Accessed 11 Jun. 2019].
8. Wikipedia (2019). *Animal source foods*. [online] En.wikipedia.org. Available at: https://en.wikipedia.org/wiki/Animal_source_foods [Accessed 11 Jun. 2019].
9. Wolfram, T. (2018). *Food Sources of 5 Important Nutrients for Vegetarians*. [online] Eatright.org. Available at: https://www.eatright.

org/food/nutrition/vegetarian-and-special-diets/food-sources-of-important-nutrients-for-vegetarians [Accessed 11 Jun. 2019].

10. Better Health (2018). *Foods from plants and animals*. [online] Betterhealth.vic.gov.au. Available at: https://www.betterhealth.vic.gov.au/health/HealthyLiving/foods-from-plants-and-animals [Accessed 11 Jun. 2019].
11. Bilić, P., Kuleš, J., Galan, A., Gomes de Pontes, L., Guillemin, N., Horvatić, A., Festa Sabes, A., Mrljak, V. and Eckersall, P. (2018). Proteomics in Veterinary Medicine and Animal Science: Neglected Scientific Opportunities with Immediate Impact. *PROTEOMICS*, [online] 18(14), p.1800047. Available at: https://onlinelibrary.wiley.com/doi/full/10.1002/pmic.201800047 [Accessed 11 Jun. 2019].
12. Gaso Sokac, D., Kovac, S. and Josic, D. (2010). *Application of Proteomics in Food Technology and FoodBiotechnology: Process Development, Quality Control andProduct Safety*. [online] research gate. Available at: https://www.researchgate.net/publication/280037787_Application_of_proteomics_in_food_technology_and_food_biotechnology_Process_development_quality_control_and_product_safety [Accessed 11 Jun. 2019].
13. Jagadeesh, D., Kannegundla, U. and. Reddy, R. (2017). *Application of proteomic tools in food quality and safety*. [online] research gate. Available at: https://www.researchgate.net/publication/317828827_Application_of_proteomic_tools_in_food_quality_and_safety [Accessed 11 Jun. 2019].
14. Toldrá, F. and M. L. Nollet, L. (2013). Proteomics in Foods. [online] Available at: https://www.springer.com/in/book/9781461456254 [Accessed 11 Jun. 2019].
15. Jane, M. (2017). *Animal vs Plant Protein - What's the Difference?*. [online] Health line. Available at: https://www.healthline.com/nutrition/animal-vs-plant-protein#section3 [Accessed 11 Jun. 2019].
16. Halkjær, J., Olsen, A., Bjerregaard, L., Deharveng, G., Tjønneland, A., Welch, A., Crowe, F., Wirfält, E., Hellstrom, V., Niravong, M., Touvier, M., Linseisen, J., Steffen, A., Ocké, M., Peeters, P., Chirlaque, M., Larrañaga, N., Ferrari, P., Contiero, P., Frasca, G., Engeset, D., Lund, E., Misirli, G., Kosti, M., Riboli, E., Slimani, N. and Bingham, S. (2009). Intake of total, animal and plant proteins, and their food sources in 10 countries in the European Prospective Investigation into Cancer and Nutrition. *European Journal of Clinical Nutrition*, [online]

63(S4), pp.S16-S36. Available at: https://www.ncbi.nlm.nih.gov/pubmed/19888272 [Accessed 11 Jun. 2019].

17. Olsen, N. and Johnson, J. (2018). *Animal vs. plant protein: What is the difference and which is best?*. [online] Medical News Today. Available at: https://www.medicalnewstoday.com/articles/322827.php [Accessed 11 Jun. 2019].
18. Pasiakos, S., Agarwal, S., Lieberman, H. and Fulgoni, V. (2015). Sources and Amounts of Animal, Dairy, and Plant Protein Intake of US Adults in 2007–2010. *Nutrients*, [online] 7(8), pp.7058-7069. Available at: https://www.ncbi.nlm.nih.gov/pmc/articles/PMC4555161/ [Accessed 11 Jun. 2019].
19. Winichagoon, P. and Chavasit, V. (2019). *INTERACTIONS: FOOD, AGRICULTURE AND ENVIRONMENT – Vol. II - Food Sources - Aree Valyasevi, Pattanee*. 2nd ed. [ebook] Nakhon Pathom, Thailand: ©Encyclopedia of Life Support Systems (EOLSS). Available at: https://www.eolss.net/Sample-Chapters/C09/E4-24-06.pdf [Accessed 11 Jun. 2019].
20. Cutillas, P. (2013). *The role of proteomics in the development of personalized cancer medicine*. [online] European Pharmaceutical Review. Available at: https://www.europeanpharmaceuticalreview.com/article/18436/the-role-of-proteomics-in-the-development-of-personalised-cancer-medicine/ [Accessed 11 Jun. 2019].
21. Edwards, A., White, M. and Cordwell, S. (2008). The Role of Proteomics in Clinical Cardiovascular Biomarker Discovery. *Molecular & Cellular Proteomics*, [online] 7(10), pp.1824-1837. Available at: https://www.mcponline.org/content/7/10/1824 [Accessed 11 Jun. 2019].
22. Keown, S. (2016). *Filling in the 'missing biology' of breast cancer*. [online] Fred Hutch. Available at: https://www.fredhutch.org/en/news/center-news/2016/05/study-critical-role-proteomics-cancer-personalized-medicine.html [Accessed 11 Jun. 2019].
23. Lippolis, R. and De Angelis, M. (2016). *Journal of Proteomics & Bioinformatics: Review Article Open Access Proteomics and Human Diseases*. [online] omics online. Available at: https://www.omicsonline.org/open-access/proteomics-and-human-diseases-jpb-1000391.php?aid=69584 [Accessed 11 Jun. 2019].
24. Aebersold, R. and Cravatt, B. (2002). Proteomics – advances, applications and the challenges that remain. *Trends in Biotechnology*, [online] 20(12), pp. s1-s2. Available at: https://www.cell.com/trends/

biotechnology/fulltext/S1471-1931(02)00206-9 [Accessed 11 Jun. 2019].

25. lumen learning (2019). *Genomics and Proteomics | Boundless Biology*. [online] Courses.lumenlearning.com. Available at: https://courses.lumenlearning.com/boundless-biology/chapter/genomics-and-proteomics/ [Accessed 11 Jun. 2019].
26. Proteomics, C. (2019). *Bioinformatics for Proteomics*. [online] Creative Proteomics. Available at: https://www.creative-proteomics.com/services/bioinformatics-for-proteomics.htm [Accessed 11 Jun. 2019].
27. Simonian, M. (2016). Proteomics in Medicine. *Journal of Data Mining in Genomics & Proteomics*, [online] 7(3). Available at: https://www.longdom.org/open-access/proteomics-in-medicine-2153-0602-1000e126.pdf [Accessed 11 Jun. 2019].
28. Utami Putri,, D., Kachidza Chiumia, F., Teng Jheng, Y. and Han, C. (2019). *SCITECH - The Role of Proteomics for Dissecting Aberrant Molecular Signaling Pathways upon Egfr-Tki Treatments in Non-Small Cell Lung Cancer - Proteomics and Bioinformatics (ISSN:2641-7561)*. [online] Scitcentral.com. Available at: http://www.scitcentral.com/article.php?journal=53&article=278&article_title=The%20Role%20of%20Proteomics%20for%20Dissecting%20Aberrant%20Molecular%20Signaling%20Pathways%20upon%20Egfr-Tki%20Treatments%20in%20Non-Small%20Cell%20Lung%20Cancer#tabs1 [Accessed 11 Jun. 2019].
29. Whiteaker, J. (2010). The Increasing Role of Mass Spectrometry in Quantitative Clinical Proteomics. *Clinical Chemistry*, [online] 56(9), pp.1373-1374. Available at: http://clinchem.aaccjnls.org/content/56/9/1373 [Accessed 11 Jun. 2019].
30. Annenberg Foundation (2017). *Rediscovering Biology - Online Textbook: Unit 2 Proteins & Proteomics*. [online] Learner.org. Available at: https://www.learner.org/courses/biology/textbook/proteo/proteo_14.html [Accessed 11 Jun. 2019].
31. Cao, Z. and Robinson, R. (2014). The role of proteomics in understanding biological mechanisms of sepsis. *PROTEOMICS - Clinical Applications*, [online] 8(1-2), pp.35-52. Available at: https://onlinelibrary.wiley.com/doi/abs/10.1002/prca.201300101 [Accessed 11 Jun. 2019].

32. Pacheco, R., Correia, S., Poeta, P., Pinto, L. and Igrejas, G. (2012). The Role of Proteomics in Elucidating Multiple Antibiotic Resistance in Salmonella and in Novel Antibacterial Discovery. *Salmonella - Distribution, Adaptation, Control Measures and Molecular Technologies*. [online] Available at: https://www.intechopen.com/books/salmonella-distribution-adaptation-control-measures-and-molecular-technologies/the-role-of-proteomics-in-elucidating-multiple-antibiotic-resistance-in-salmonella-and-in-novel-anti [Accessed 11 Jun. 2019].
33. Proteomic.org. (2004). *Proteomics?*. [online] Available at: http://www.proteomic.org/html/proteomics_.html [Accessed 11 Jun. 2019].
34. Wang, K., Yang, Z., Qing, D., Ren, F., Liu, S., Zheng, Q., Liu, J., Zhang, W., Dai, C., Wu, M., Chehab, E., Braam, J. and Li, N. (2018). Quantitative and functional posttranslational modification proteomics reveals that TREPH1 plays a role in plant touch-delayed bolting. *Proceedings of the National Academy of Sciences*, [online] 115(43), pp. E10265-E10274. Available at: https://www.pnas.org/content/115/43/E10265 [Accessed 11 Jun. 2019].

5

Proteomics and Aquaculture

CONTENTS

Proteomics deals with the study of proteins in food products and the various natural organisms. The study reveals various outcomes and the components of the food products. This study helps us to know about the nutritional value of the food in terms of the proteins in it. Proteins are key elements that build up the muscle mass of a body. Hence, proteomics can help uncover several aspects and find various results from the researches that can help in the human health and wellbeing. This chapter brings the focus of the readers to the application of proteomics in aquaculture products.

5.1 INTRODUCTION

Consumers must certainly be aware of the ingredients of the food they are consuming. This is the duty of the producers to inform the consumers, without intentionally misleading them, about the content of the food products they are buying and taking home for consumption and update them about the source of their products..

However, as the demand for food across the globe is increasing and the trade regarding the food products is on the rise, the risks regarding the adulterations in food products taking place anywhere in the entire food chain, go up.

The food products can be replaced intentionally, in entirety or in a partial way, by the products that are similar in nature, but much poorer in quality and lesser in price. They can also be substituted in an unintentional way such as in the case of wrong labeling of a product. Both such instances of commercial fraud may impact not only the consumers of the product but the food industry.

Figure 5.1: The produce from an aquaculture habitat or field.

Source: https://farm1.staticflickr.com/120/294961236_5d6032d1b4_b.jpg

Some instance of the frauds that take place, in general, are substituting the main ingredients with some cheaper counterparts of the product, wrong labelling of the product in terms of the animal species used to produce it, wrong labelling of the product in terms of the proportions of the ingredients used in it, providing wrong information about the product to the customers such as selling the aquaculture fish as wild-caught and wrong labelling of the synthetic foods as organic ones.

The adulterations in the food products make for a big issue that concern not only the economic aspects of a sold product or the fraud related to it, but also focuses on the safety of the product. There may be instances where the food ingredients of a food product are not declared as toxic or the ones that may cause allergy to people. These kinds of instances may cause grave harm to the consumer's health, which poses as a medium for a possible form of risk for the health of the public.

In addition to this, wrong labeling of a product may have great impacts on the eating habits of some kind of people who may not like to have certain foods on the grounds of religious inclinations, self-restraint or some choices that may concern the lifestyle of people like being vegetarian.

For the purpose of countering the various adulterations and mixings in the products, several laws have been passed that require the producers to publish or declare complete and truthful information about the product that is in the market for consumption.

Considering the reasons and instances stated above, the authentication of the claims that concern the labeling of food, must be guaranteed and to accomplish this, there is a need of dependable and precise methods of analysis. These methods analyze the food products to confirm whether the ingredients that are in use in the food products are in accordance with the one that is being demanded by the consumer and matches the declaration made by the producer or the seller.

There are several different techniques for this purpose of food authentication, for example:

- HPLC
- Stable isotope ratio (SIR) analysis
- Genomics
- Proteomics
- Metabolomics

- NMR spectroscopy and
- Trace element analysis.

5.1.1 Shellfish

The products in the category of seafood, that involve shellfish and fish, fall under the most widely bought and sold food products at the global level. Due to the great phenotypic similarity among the species that are interrelated with each other, there may be a presence and occurrence of various adulterations taking place in the food products that may both be intentional and unintentional.

Figure 5.2: Shellfish have a lot of applications in the product preparation in the aquaculture and fisheries.

Source: https://cdn.pixabay.com/photo/2019/02/13/06/32/cockle-3993785_960_720.jpg

The adulterations may be among the ones such as replacing a very valuable and costly species with a species that is very similar in nature but is far less valuable than the one being replaced or the product that has been derived from aquaculture may be termed as wild fished one.

Hence, in order to counter this adulteration and make sure that the genuineness of the product is maintained, several laws have been formed to deal with the trading of these products.

For instance, in Europe, the laws require the manufacturers to label the seafood products by making sure that they indicate the proper commercial recognition of the species used, the method adopted in the production which may be among farmed or wild, and include the geographical region in which the particular sea creature (fish or shellfish), was farmed or caught.

Hence, to analyze the requirements of such laws and indications, there must be an adoption of a suitable analytical approach. In a research conducted by An and others, it was found that profiles relating to electrophoretic protein can be utilized to distinguish between the species that are extremely closely related.

Even in the case where the study was limited to only three shrimp species and where there was no knowledge of the sampling available from before, they were capable of distinguishing between them making use of SDS-PAGE or IEF of the proteins that were obtained from the samples of muscle.

The use of these types of gels has also been made to distinguish among the shrimps, crab meat, lobster, cephalopod meat and the fishes. Nevertheless, all the research conducted before, it was validating that there wasn't any specific or single way to extract or a single electrophoretic system which is applicable to all the systems.

The SDS-PAGE and IEF gels that were composed of the sarcoplasmic proteins and were soluble in water, had good results while distinguishing the raw samples. On the other hand, the SDS samples worked in a better way while dealing with the cooked muscle.

In addition to this, the sampling of the gels was not done in an extensive way, while the gels were analyzed from the visual capabilities, without making use of any statistical analysis. For the purpose of distinguishing between the species of prawn and shrimp that are used for commercial purposes, the muscular proteins that are soluble in water were used. This was done making use of native IEF that is also identified by the proteins particular to the species, by LC-MS/MS.

This investigation represents the most complete electrophoretic research portrayed to date for the recognition of closely related species since the shrimp and prawn species contemplated were unambiguously recognized, and various populations were incorporated for a few of the species, thus evaluating intraspecies fluctuation.

The pI of the marker protein groups was accomplished utilizing image analysis software and consequently, the outcomes can be utilized as a kind of perspective database for the distinguishing proof of obscure examples.

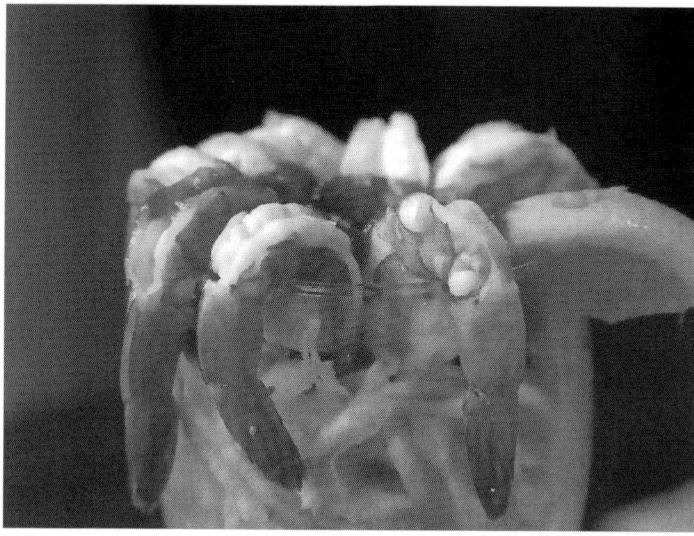

Figure 5.3: A shrimp cocktail prepared for immediate refreshment and prolonged nutrition.

Source: https://upload.wikimedia.org/wikipedia/commons/f/f8/Cocktail_1_bg_060702.jpg

No cooked examples were analyzed, yet the way that the species-specific proteins were recognized as heat-stable sarcoplasmic calcium-binding proteins (SCPs) increases the potential utilization of the strategy in heat processed and prepared to-eat items, even though it presently can't seem to be proven. Additionally, SCPs have been built up as shellfish allergens, so an approach focusing on this brand of proteins would affect not only the authenticity of food but also its safety.

2-DE has additionally been utilized for examining quantitative distinctions between two marine mussel species. A few protein spots (n = 37) were observed to be differentially communicated by the computer-based analysis of silver-stained 2-DE gels, albeit just 15 of them were recognized by methods for PMF or PFF research.

Nevertheless, the researchers noticed that the distinctions observed could be brought about by the changes in the environment and also the genetic inconstancy since every species came into existence from a different region.

In an ensuing report, similar researchers detailed species-specific peptides for the three European marine mussel species by joining 2-DE protein segregation, PMF investigation of projecting protein spots and item particle peptide *de novo* sequencing.

The peptides could likewise be recognized utilizing SIM, which is known to improve affectability though can be lacking specificity in certain examples. Although just four biological replicas from every one of the three species were considered, the investigation represented the main use of a total bottom-up proteomics way to deal with species authentication.

5.1.2 Fish

Concerning shellfish, traditional electrophoretic strategies, for example, SDS–PAGE and IEF have been widely utilized for fish species authentication. In an interlaboratory study completed by nine research facilities, the way in which the optimized SDS–PAGE and urea-IEF standard operation methods performed, was assessed for the ID of ten commercial fish species after they had been cooked.

It was stated that the two techniques performed diversely according to the species, and thus the utilization of the two strategies was advised for the distinction in the species. The requirement for legitimate reference examples to keep running in parallel to the unknown example being investigated is one of the principal issues related to electrophoretic strategies.

Figure 5.4: Fish is excessively used in the preparation of aquaculture products and are rich in proteins.

Source: https://live.staticflickr.com/7491/16047558042_a64bdae9dd_b.jpg

Interspecies inconstancy was evaluated for a few of the species by investigating up to three examples, although no data about various populations or geographic areas were accounted for. The researchers presumed that intraspecies similarity was bigger than interspecies similitude, however, it must be seen that this could be because of examples from one species being taken from the equivalent geographic region.

Renon and others utilized the IEF sarcoplasmic protein example to differentiate between swordfish, blue marlin and Mediterranean spearfish, in new, frozen, and cold-smoked filets, however, just a single biological replica for swordfish and marlin, and two for spearfish, were utilized.

Since the 1990s, differences in the subjective 2-DE profiles of water-dissolvable muscular proteins have been utilized for the distinction between fish species, for example, gadoid fishes, flatfishes and hake species.

In these researches, a few examples from far away fishing banks were researched for a few of the species to evaluate intraspecies dissimilarities, and three to four biological replicas for every species or topographical area were considered.

5.2 THE USES OF PROTEOMICS TO ASSESS THE EFFECTS OF CLIMATE CHANGE ON THE MARINE PRODUCTS

In the subject of marine biology, proteomics helps to understand the systems that form the basis for the changes in the metabolism of invertebrates and fishes, that may be brought about by the changes that may be catering to the biotic and abiotic factors. However, there are many types of research that are going on currently which are being done under the laboratory conditions.

There is also a description of the uses that may be thought of as the illustrations of the possible role that proteomics can play in the evaluation of the impacts of climate change on the organisms found in the marine ecosystem.

Figure 5.5: Some of the enhanced applications of algae from aquaculture.

Source: http://res.publicdomainfiles.com/pdf_view/155/13987633829662.jpg

5.2.1 Environmental Applications

There are two perspectives from which the applications of proteomics in the toxicology concerned with the environment, have been developed. These may be given as:

- The comprehension of the new targets at molecular levels that may be associated with the toxic compounds so that the way they undertake different activities can be understood properly
- The way the information, generated by the alterations in the way the proteins are laid out, is used so that the activities of the toxic substances can be screened, and the forecasts can be made

There is an explanation of the uses of proteomics in the assessment of the response of the marine organisms at their cellular level on being subjected to a mixture that is toxic in nature, in a research that has been conducted recently.

Regardless of the fact that the uses of proteomics in the marine science have not been as many as that in the field of biomedicine, the procedures in the proteomics and the strategies related to it, have been utilized in the researches and studies that have been conducted in correspondence to the marine ecosystem.

This has majorly been done to assess the various biomarkers that correspond specifically to the protein found naturally in the environment and to define the various stress responses related to it (Amelina, Apraiz, Sun, & Cristobal, 2007; Nesatyy, Ammann, Rutishauser, & Suter, 2006; Nesatyy & Suter, 2008; Prunet, Cairns, Winberg, & Pottinger, 2008; Sheehan, Tyther, Dowling, & McDonagh, 2007; Viarengo, 2006).

With regard to this, proteomics has been utilized in a way so that the changes in the proteins such as metallothionein, cytochrome p450, cholinesterase, monooxygenase system and others that are in accordance with the problems in ecotoxicology can be assessed. The issues may be in a wide range of marine species that act as watch guards and the assessment may act as a great indicator for the stress on ecosystem and the probable threats to the humans in terms of their health.

There has been an introduction of a new discipline, keeping in mind, the applications of proteomics in the assessment as mentioned above, which goes by the name of Ecotoxic proteomics, as stated by Gomiero and others in 2006. It is important to note that the "sentinel species" are bivalves when seen from the perspective of seafood that majorly comprises of mussels and clams.

The factors that indicate that they may be the best choices to deal with marine pollution may be among the large amounts in which they are found, their abundance in the various geographical parts, their taxonomic diversity and their potential to be in close vicinity to the sources of ecological toxicology that re aquatic in nature. This is because they have great capability to adapt to the pollution currently there in the environment.

In addition, there is the presence of some reports that depict the assessment of the effects in the environment that have been observed in a few species of aquatic animals, mostly the aquatic vertebrate which involve Atlantic Salmon (*Salmo salar*), rainbow trout (*Onchorynchus mikiss*), flatfish (*Limanda limanda*), zebra fish, cod (*Gadus morhua*) and a few similar other species, which may be put under the category of the possible sentinel species.

The study of the impact on species that have been listed here, involves the use of proteomics tools. The main point that may be given and emphasized over here is that the in almost all the situations and scenarios, the research studies on environment can provide the results concerning the safety aspects in seafood.

5.2.2 Applications in Seafood

The field of proteomics is based on genomics in a very peculiar way. It is found in a way in which, if there is a shortage of knowledge about the specific species in the genomic form, it can result in extremely restricted achievement in the identification of protein in that species (Graham, Elliott, & Van Eyk, 2005).

There is a very less amount of information in the databases that can be related to the shellfish and marine fish, which in turn makes it a little more tedious to study and get the information on the proteomics concerning them (Piñeiro, Barros-Velázquez, Vázquez, Figueras, & Gallardo, 2003).

The only kinds of marine and shellfish on which the information is available, are estuarine fish (*Fundulus heteroclitus*), zebra fish (*Danio rerio*) and Japanese pufferfish (*Takifugu rubripes*). The availability of information is in the form of a complete genome sequence, which makes it easy to apply the findings of the proteomics research to the aquatic ecosystem and the creatures dwelling in it.

The research findings of the proteomics studies have been applied in advance to the fields concerning some particular elements of the research and development concerning seafood. Though the impacts of climate change, that may be seen using the approaches involving proteomics, on the seafood, that is being prepared, has not been found in any of the compilations of reviews or researches accessed till date.

In view on the information that is available in the scientific form, the applications of the proteomics regarding the aspects of seafood and its production and the safety concern with it, have been compiled in various researches and studies and put down in three principal parts:

- The studies on food safety
- The applications of the authentication and taxonomy
- The nutritional side of the seafood

5.2.3 Applications Concerning the Safety of Seafood

There are various forms of issues surrounding the safety of seafood. These may be listed as:

- Pathological
- Allergenic
- Parasitological

There are various tools and methods designed for authentication of taxonomic for gadoid species and it is developed by using different types of proteomics methodologies and a specific fast analytical procedure has been patented in Spain (patent Spanish number 200603287 by Carrera and others, 2008).

The presence or absence of specific peptides, which is identified based on analyzing their masses by using mass spectrometry in fresh, processed or frozen fish. The use of proteomics has been widely used for taxonomic purposes and now extended to other marine species such as Perciformes and Pleuronectiformes (flatfishes), as seabasses, mullets, seabreams and more recently to shrimp and prawns

The primary scientific objective in engaging in nutrition research activities is to assess the role of the diet in metabolic regulation to ensure the safety of human health. Depending on individual genotypes, nutrients may be used in altering metabolic pathways through continuous monitoring of the functioning of genes and change in the types of proteins, and thus improving the health status.

Nutriproteomics or Nutritional proteomics means the application of the proteomics methodologies in conducting the nutrition-based studies and, among other topics, focus on the relationship between proteins and interaction of bioactive food ingredients, which is analyzed and examined by using two different approaches:

- Identifying the effect of bioactive food ingredients on protein synthesis with the help of gene expression method
- Determining the interference of these ingredients with other proteins and looking for alteration and modification or for protein/molecule interaction.

As it was done by using omics-based nutrition approaches (nutrimetabolomics and nutritranscriptomics), the majority of investigations of nutriproteomics have been performed using animal models (Wang and others, 2006).

It is worth noticing that there are only a few applications of nutriproteomics to seafood in order to analyze the bioactive ingredients.

Learning of the molecular mechanisms is one type of application that is regulated by the consumption of PUFA. Proteomics is basically used to identify proteins in order to identify the alteration of serum levels by the administration of long-chain x-3 polyunsaturated fatty acids (LC-PUFA) from fish oil.

It is possible to do the comparison with the alterations produced when sunflower oil, having a rich source of oleic acid, was distributed. A down-regulation of the serum proteins that is responsible for causing coronary heart disease among patients who consume food that has rich source of fish oils, was evident.

Proteomics applications to the research of nutritionally relevant proteins have been determined, and examining their role in human health. In that context, fish and various other marine products containing nutritional, necessary minerals and relevant proteins, related to organoleptic properties of fish products and consequently to the grade of their quality, with a higher proportion of essential amino acid, which are available in excess quantity in the edible parts.

There is a higher possibility that the structure of these proteins affected during the storage and manufacturing process (Herrero, 2008).

Proteomics methodologies play a very important role in examining the changes that occurre in proteins during tuna and cod frozen storage. Monitoring the variation in nutrition values between the farmed and wild species is also a major applications of nutriproteomics.

Researchers are giving special emphasis to cultivated species as an adaptation to global climate change. Proteomics strategies play an important role in preliminary biochemical and physiological investigations. The relationship between sarcoplasmic proteins and cold acclimation expression in carp muscle have been studied.

When fishes were cooled at a temperature below the 10 degree Celsius, there was a strong collection of fragments of the enzyme creatine kinase (CK), consistent with the observed higher muscle CK proteolysis associated with the decrease of water temperature.

5.3 PROTEOMICS TO EVALUATE QUALITY OF AQUACULTURE SPECIES

5.3.1 Proteomics in Seafood Quality

Quality assessment in aquaculture majorly considers various types of factors such as

- Geographical origin, authentication and traceability;

- Storage and processing;
- Production processes;
- Functionality sensory;
- Nutritional qualities.

There is significant progress witnessed in this area of research corresponding to the various products from aquaculture. Fish products are becoming more popular and more research and studies are conducted to assess the quality in comparison to food products from mollusks and crustaceans. In various proteomics studies, all these key differences are rejected.

As far as the identification of food products is concerned, classical methods were not effective. For example, in case of differentiating among *Penaeidea* shrimp species having phenotypic similarities and regular elimination of external carapace during the processing process.

Proteomics approaches based on selected tandem mass spectrometry ion monitoring (SMIM) or 2-DE were used as alternative authentication methodologies for the authentication and identification of relevant shrimp species that are commercialized.

The classification of species-specific proteins and peptides confirmed the accuracy of these approaches for authentication of seafood product allowing the different characteristics of several commercial, almost similar, shrimp species.

It was observed that the existence of allergenic proteins in crustacean food products plays a major role in deteriorating the quality of food. Protein tropomyosin is one of the major allergens in crustacean food. This type of protein immediately reacts with the human body IgG/IgE and weaken the immune system of a person that generates allergenic symptoms.

Proteomic profiling of the allergen tropomyosin with specific monoclonal antibodies play an important role in checking the presence and allergenic potential of the protein in mollusks and shrimp. The result of the study showed that when the shellfish meat when exposed to high temperature , it enhances recognition of multiple variants of tropomyosin in the analyzed shellfish species, indicating a possible alteration of allergenicity potential because of heat treatment.

In contrast, in a protein profiling study, it was observed that the high pressure steaming significantly decrease the levels of tropomyosin in shrimp (*P. monodon*) primarily due to the degradation of protein, thus minimizing the allergenicity of tropomyosin as compared to other heat treatments methods.

The gel-based method is a type of method that is used to determine the quality of protein and digestibility in squid fillet that was treated with different drying methods and high hydrostatic pressure methods (HHP). HHP treatments help in altering the functional properties and proteins of squid and reduced the allergenicity of tropomyosin Tod p1 by secondary structure and protein unfolding modification, thus providing the opportunity for eradicating allergenicity of squid.

Complementary proteomics and biochemical methods were used to classify the chymotrypsin from the midgut gland of *Penaeus californiensis*, yellow-leg shrimp for food processing, biotechnological applications, and control of post-mortem muscle melanization and softening.

5.3.2 Proteomics in Fish Quality

Fish quality is mainly concerned with "fish-eating quality" that is available to humans for consumption and can be affected by a number of factors that have been analyzed by using proteomics technologies. All these factors are categorized into various types, such as:

- Fish development phases and production stages, that is, during the ante-mortem period (influenced by diseases, genetics, nutrition, stress, temperature, and slaughter method); and
- Fish processing techniques, that is, during the post-mortem period (influenced by storage temperature, time and processing techniques).

Also, one of the approaches that are concerned with the quality is the traceability (origin) and authentication of fish products. Proteomics approaches are very useful and play an important role in farmed fish as it is used as a monitoring tool to understand the biological mechanisms related to changes in quality, to produce farmed fish of high quality and to ensure the safety of humans as well as environment.

During the antemortem period, fish welfare can play a major role in fish quality treats. Fish welfare is generally considered as a complex concept that needs a holistic and multidisciplinary approach in its studies. Welfare is also many times strictly related to stress.

Nevertheless, stress responses are not capable enough to give all the relevant information about fish welfare, because in aquaculture, this last one is primarily related with tertiary effects of stress response that are generally indicative of repeated, prolonged, or unavoidable stress that is closely associated to maladaptive effects on reproductive function, growth, disease resistance and immune function.

5.4 PROTEOMICS TO ASSESS THE POTENTIAL RISKS ASSOCIATED TO AQUACULTURE

5.4.1 Proteomics in Seafood Safety

Seafood safety is primarily concerned with preventing and minimizing the risks associated with the consumption of seafood. It was observed that with the increase in the consumption of seafood, the chances of seafood-related health risks increased.

Some of the seafood borne health risks are infections mainly caused by bacteria, virus and parasites, or risks associated with the supplying of pollutants or toxins through the food chain. According to a recent survey, it was found the risks from the contaminated seafood is almost same in both the developing as well as developed countries, however, in countries with a proper health care facility, some of these risks can be minimized.

Mollusks are filter-feeding species and their daily nutrition requires filtration of several liters of water per day. Therefore, gathering of various types of pollutants such as metals, toxins, or chemicals has been witnessed.

Initially, proteomics was widely used for environmental assessment and continuous monitoring of marine pollution. The changes in protein expression from sentinel organisms residing in the monitoring areas provide both spatial and temporal information about the water quality.

The identification and detection of bacterial pathogens that are usually found in seafood is one of the major objectives to ensure the safety of seafood and improving the quality of seafood.

MALDI-TOF MS analyses make the job of researchers easy in obtaining reproducible mass spectral fingerprints of the gram-negative bacteria, *Vibrio parahaemolyticus* that is primarily responsible for making seafood infected and that cause a major threat to the health of the human body. The collected fingerprints can be helpful to discriminate between the phylogenetically diverse *Vibrio* strains.

Finally, with the increase in antibiotic resistance, the global threat to the human as well as the environment has increased. The increase in the quantity of antibiotics in aquaculture has caused these antibiotics to come in contact with the environment.

The aquatic environment is therefore considered as an important reservoir of antibiotic resistance genes (ARGs) and is responsible for the dissemination of antibiotics in various areas as well as to human pathogens.

Metagenomics has been applied in order to establish a link between the resistive and the metagenomics data. Therefore, the transfer of antibiotic resistance from natural microbiota to the human body may pose a serious threat to humans.

Various methods to monitor the quality of water has been applied to marine sediment, which is close to aquaculture farms and it comes under notice that the native resistome has been enriched by antibiotic use at the fish farm.

However, there are only a few genes that are observed by this method. A metaproteomic analysis is a type of method that can address this problem as it is applicable to all types of species, on the entire protein complement of an environmental sample at a given time.

The application of this methodology in freshwater sediments on bacterial communities has revealed very useful information in the adaptation of the community structure to high levels of metal. Finally, metaproteomic coupled to steady isotope probing (SIP) from environment and climatic conditions enable the characterization of microbial assemblage through quantifying and identifying the enrichment of isotopically labeled proteins.

Environmental metaproteomic provides a lot of useful ways and methods for improving the understanding and management of microbial ecosystems and their global impacts.

5.4.2 Proteomics in Fish Safety

Fish safety is concerned with ensuring that fish is free from various types of diseases caused either by fraudulent practices such as fish contamination, fish diseases, fish substitution and fish allergens.

There are various international bodies responsible for ensuring the safety of fish and imposed various rules and regulations by keeping in mind the safety of consumers and the organizations that engage in the production of fishery products need to comply with the same.

Among those, European Union, Directives EC No 2073/2005 focus on the regulation of microbiological contamination and EC No 104/2000 and 2065/2001 regulate labeling characteristics.

One of the most important steps taken so far is to ensure that the labeling should be included in every seafood product and it must include the scientific species name, commercial designation, geographical origin, and production method (farmed or wild).

In order to gain more insight into this matter, suitable detection methods must be put into place. Proteomics is very useful for researchers in providing a reliable, accurate, delicate, and specific control method. In the case of fish authentication, there are various studies already available reporting the use of proteomics-based methods.

The methodology explained in this chapter was validated in 10 commercial hake products and also provides various alternative ways for fish parvalbumin detection. In a similar manner, a researcher successfully applied proteomics that helps in identifying 25 different types of fish species from three different orders describing specific PRVB peptide masses for each species.

More recently, a new approach known as shotgun proteomics approach was used to identify the edible muscle of gilthead seabream, which revealed quantitative differences in PRVB between wild and farm animals, with this protein being overexpressed in farmed fish samples.

5.5 FUTURE PERSPECTIVE

Proteomics-based strategies play an important role to offer a holistic view of welfare, security, safety, and traceability of species of interests for aquaculture. However, it is important to note that for species of interest to aquaculture, only a few fully sequenced genomes are currently available.

Recent technological development and advanced approaches in the genomics have been directly related to the growth of high-throughput sequencing facilities and minimization of sequencing costs. Therefore, it can be assumed that research on shotgun proteomic along with genomic sequencing attempts will help in getting an insight into aquaculture species.

In particular, the formation of non-invasive methodologies using serum or fish plasma and seafood hemolymph that can link high-throughput and high-resolution methods to keep a check on diverse parameters related with safety and/or quality is a direction of high interest.

Recently, with the increase in understanding of the earth microbiome, it is becoming more accessible. There are various new methods and approaches that helps in increasing the understanding of the dynamics of the microbial communities from sediments, water, and gut microbial communities.

All these methods are considered as a very useful tool to evaluate aquaculture, security, safety, and welfare. Proteomic research in aquaculture is undertaken, along with ensuring the sustainability of the environment and play a very important role in seeking methodologies that will ensure the development of sector management strategies and contribute to preserving the quality of the environment, thus protecting human health.

REFERENCES

1. Campos, A. and de Almeida, A. (2016). Top-Down Proteomics and Farm Animal and Aquatic Sciences. *Proteomes*, [online] 4(4), p.38. Available at: https://pdfs.semanticscholar.org/68c5/fb560c1c208fcd8ae3c6229063e8bee37ea6.pdf [Accessed 10 Jun. 2019].
2. Ortea, I., O'Connor, G. and Maquet, A. (2016). Review on proteomics for food authentication. *Journal of Proteomics*, [online] 147, pp.212-225. Available at: https://www.sciencedirect.com/science/article/pii/S1874391916302792 [Accessed 10 Jun. 2019].
3. Piñeiro, C., Cañas, B. and Carrera, M. (2010). The role of proteomics in the study of the influence of climate change on seafood products. *Food Research International*, [online] 43(7), pp.1791-1802. Available at: https://ucanr.edu/datastoreFiles/608-153.pdf [Accessed 10 Jun. 2019].
4. Rodrigues, P., Campos, A., Kuruvilla, J., Schrama, D. and Cristobal, S. (2017). Proteomics in Aquaculture. *Proteomics in Food Science*, [online] pp.279-295. Available at: https://www.researchgate.net/publication/316344446_Proteomics_in_Aquaculture [Accessed 10 Jun. 2019].
5. Tedesco, S., Mullen, W. and Cristobal, S. (2014). High-Throughput Proteomics: A New Tool for Quality and Safety in Fishery Products. *Current Protein & Peptide Science*, [online] 15(2), pp.118-133. Available at: https://pdfs.semanticscholar.org/3149/a54ff7c42dfd569ce1db908e47d7c5edf4b4.pdf [Accessed 10 Jun. 2019].

6

Proteomics and Food Processing Technologies

CONTENTS

Proteomics is basically the study of proteins with respect to its structures and functions. Proteomics can be divided into three types. This chapter explains about proteomics and its types. Broadly, there are three types of technologies used in food processing. They are named as thermal, non-thermal and other food processing. This chapter includes several technologies that are used in this era for preservation and processing of food. Thermal processing technologies include microwave heating, radiofrequency heating, infrared heating, ohmic heating etc. Non-thermal food processing technologies include high hydrostatic pressure processing, irradiation, ultraviolet light etc. Other food processing technologies include fermentation, extrusion, baking etc. in this chapter, all of these technologies are thoroughly explained.

6.1. INTRODUCTION

The study of proteins, specifically about their structure and function is referred to as proteomics. With respect to different stress conditions, there are changes in metabolism. Proteomics is the methodology to study those changes. Different type of techniques for protein analysis includes MALDI-TOF/MS and 2-D gel electrophoresis etc. These techniques are also very useful in the treatment of disease.

Protein is referred to as highly complex substances which is present in all living organisms. Proteins are polymers of amino acids. Proteins act as a key factor in metabolic activities. The sequence of amino acids which are encrypted by mRNA determines the primary structure of a protein.

In the secondary structure, one type is alpha helix. Alpha helix is a region of the polypeptide, which is foldable into a corkscrew shape. The linear structure of polypeptides is a beta structure which bonds together in order to form a flat beta sheet. Interaction between turns and coils take place so that the formation of unique 3-D shape takes place and so creates a 3-D structure of the final protein.

In order to make the final active protein, a number of proteins have a variety of different polypeptide subunits. In these proteins, the different subunits of polypeptide interact so that the quaternary structure can be formed. By studying the human genes and proteins, a new development has taken place which is the recognition of new drugs to treat several diseases. In order to identify the proteins which are related to any disease, genome and proteome information is used.

Food processing can be defined as the conversion of raw materials into foodstuffs or edible food products by apply different technologies and scientific knowledge. There are a number of processes and techniques which can be used to convert easily spoilt, bulky and also uneatable food into concentrated, delicious and shelf-stable food.

Sometimes, some changes in the food products reduce the time take to prepare or cook the food. Many times, food processing enhances the product obtained after cooking by increasing, deliciousness, portability, convenience and storability.

When the food processing is done, the professional performing the same should have knowledge about the general characteristics of raw food materials, processing factors such as quality, principles of food preservation, water and waste management, packaging and procedures for proper sanitation.

An essential link in the food chain is the technologies used for food processing. There are a number of food processing technologies and thus, there are changes in popularity which depend upon the change in consumption patterns and product popularity. The new innovations in these technologies have provided additional advantages.

6.2. PROTEOMICS

The complete group of proteins of an organism which is encrypted by the genome is termed as proteome. The total protein contained in a cell or an organism is referred to as proteomics. During different stages of the life cycle or under stress conditions, the proteomics helps in better understanding the changes in the protein expression.

Proteomics helps in understanding the structure and function of different proteins. In addition to this, proteomics also helps in understanding the interactions taking place between protein – protein in an organism. Using proteomics, detection of small defects in protein structure and function, as well as changes in the process, can also be done. With respect to drug development and biological processes, the proteomics is important.

Proteomes are the set of proteins that makes up the cells as well as tissues of any living organism. Likewise, genomes are the set of inherited genes that are present in each individual that includes genes that control protein synthesis in the body.

The human body has around 30,000 genes as well as several hundreds of thousands or millions of different proteins. Identification of all the genes has been done. But science has not identified all the proteins present in the human body. Only around 80,000 proteins have been identified till now.

In order to determine a pathological problem and a therapeutic solution, the quantitative or qualitative change of a single type of protein is enough. This is what functional biology is.

In 1995, the term 'proteomics' was first originated. It was defined as the representation of the entire protein complement of a cell, tissue or organism. Proteomics is particularly the study of proteins. This refers to the study of composition, functions, structure and interactions of different kind of proteins that conducts the activities of the cell.

Genomics is the study that gives a rough estimation of the expression of a protein. Most of the proteins work together with other proteins and proteomics aims to identify which protein is interacting.

Most of the time it is considered that the proteomics id the advanced step in the study of biological systems. In comparison to genomics, proteomics is very much complicated. This is said to be true because the genome in any organism is constant, but the protein expression keeps on changing with several conditions such as time, micro and macro environmental conditions.

Proteomics aims to find comprehensive differences in the group of protein. Thus, it will help in establishing measurable criteria that are used in order to define a disease, therapeutic targets and precision or personalized treatments. Since several drugs work by modulating the function of proteins, so its value can be found in the therapeutic and diagnostic ability.

Thus, the proteomic analysis works in a direction to identify the differences within samples at the protein level with respect to levels of abundance, structure or function. Proteomic analysis is also data-driven like some other analysis. It is able to obtain impartial protein profiles for a variety of extremities that can deviate, for example treated and untreated cell models or healthy and infected tissue.

After the proteolytic processing of the parent proteomes, proteomic data can be obtained. The extracted peptides are analyzed using the instruments coupling nano- liquid chromatography to mass spectrometry. These instruments can obtain several spectra of peptides and further fragmentation to mass spectrometry spectra.

After fragmentation, the fragmented peptides produce a unique mass spectrometry spectrum which then is compared to a theoretical spectrum of amino acids which is further can be defined through the repositories of public genes. The peptide sequence matches are then mathematically deduced. Using that data, the proteins can be identified. From such analyses, peptide markers can be used as surrogates for the gene products from which they are derived.

By analysis of these peptide markers using the differential expression, proteomics can help in identifying the gene products that define a phenotype. There are a number of label and label-free quantitative analysis that is present in order to compare and differentiate the quantities of proteins in the starting samples. In order to establish the interpretation of the proteome regulation between the samples of different phenotypes, this process can be done.

Proteomics can be categorized into three kinds: expression, structural and functional proteomics.

Expression proteomics

In order to study the qualitative and quantitative expression of the total protein contained in an organism under two different conditions, the expression proteomics can be used. For example, for understanding the protein that is accountable for the infected state of the cell, the comparison should be done between a normal cell and treated or infected cell.

In order to examine the expression of protein patterns in any abnormal cell, the expression proteomics can be used. For example, a comparison of tumor tissue sample and normal tissue can be done, in order to analyze the differential protein expression. For observing the changes occurring in the protein expression, which may or may not be present in the tumor tissue, comparison is done and for that several techniques are used such as 2-D gel electrophoresis, mass spectrometry techniques.

While identifying these proteins, it will provide valuable information about the molecular biology behind the tumor formation.

Structural proteomics

For a better understanding of 3-D shape and the structure of functional proteins, the structural proteomics is used. The prediction of the structure of the protein is done when its amino acid sequence is provided by sequencing or from the gene using a method named as homology modeling.

Structural proteomics is used to provide thorough detailing about the structure and function of the protein complexes which are present in a cellular organelle specifically. By using the structural proteomics, it is possible to determine all the proteins that are present in any complex system, for example, ribosomes, cell organelles and membranes. And, by using this proteomics, characterization of all the interaction occurring between proteins and between protein and protein complexes.

For determining the structure, different technologies are present namely, X-ray crystallography and NMR spectroscopy, etc.

Functional proteomics

For understanding the function of protein along with the molecular mechanisms taking place within the cell which further depends upon the identification of the proteins that are interacting, functional proteomics is used. The interaction of any unknown protein with its partner proteins that belong to a particular protein complex involving any specific mechanism, is proved to indicate the biological function of that protein.

By explaining the interaction occurring between protein and protein, a detailed explanation of the cellular signaling pathways can be illustrated.

6.3. FOOD PROCESSING TECHNOLOGIES

There is a number of technologies that are utilized by the food industry in order to preserve the food material. Examples of such technologies can be thermal processing, refrigeration, freezing and dehydration, etc. these food preservation techniques are used in order to increase the shelf life of the food by removing harmful pathogens that are present in the food and thus, minimizing microorganism and enzyme that causes spoilage.

In order to achieve the extension in shelf life and preserve quality, a general concept is associated with food processing. This concept includes:

- addition of heat,
- removal of heat,
- removal of moisture, and
- packaging of foods in order to maintain the desirable characteristics that are established through food processing.

6.3.1. Thermal Processing Technologies

Microwave heating

Microwaves are the electromagnetic radiation that have a frequency ranging from 300 MHz to 300 GHz. The heat generated by absorbing the microwaves by a dielectric material which eventually results in the microwaves providing their energy to the material along with a relative increase in the temperature is referred to as microwave heating.

In microwave heating, the electromagnetic field interacts with the whole food and provide bulk heating. The microwave heating has become a field of interest for scientists as well as consumers because of its capability of bulk heating, increasing temperature rapidly and easy to clean.

The capability of bulk or volumetric heating, microwaves can minimize the total time taken for heating. Thus, it will also reduce the overall severity of the process and so, it will lead to greater preservation of the desired quality characteristics of the product.

In comparison to conventional heating processes, the time required to reach the targeted temperature by using microwaves is one-quarter of the time required by the conventional heating processes. Microwave technology is flexible enough to provide ease of turning on and off. It is also responsive while batch processing.

For processing food with microwaves, there are some major applications of microwave processing such as pasteurization, tempering, blanching, sterilization, cooking and dehydration. Some other uses of microwave technology include coagulation, coating, baking, puffing, gelatinization and roasting. The microwave pasteurization operates at the temperature ranging from 60°C to 82°C whereas microwave sterilization operates at a temperature ranging from 110°C to130 °C.

Figure 6.1: Microwave oven.

Source: https://commons.wikimedia.org/wiki/File:Weibolu.jpg

Microwave oven using the technology of microwave heating is a very common home appliance. The heating done by microwaves is now used widely in the food industry in a number of applications.

Microwave processing is now commercially used in several applications such as bacon precooking, meat sausage cooking and baking products for examples fish blocks, frozen potato puree and beef meats. It is also used in pasteurizing various products such as beef frankfurters, mashed potatoes, packaged acidified vegetables, orange, apple cider, in-shell eggs and raw meats.

Infrared heating

Infrared heating is heating done by electromagnetic radiation which is generated by using a source such as quartz tube, metal rod and quartz lamp which results from the vibrational and rotational energy of the molecules. Generation of thermal energy is done because of the absorption of radiating energy. Infrared heating includes some basic characteristics such as high heat transfer capacity, fast process control, no heating up of air present in the surrounding and penetration of heat directly into the product.

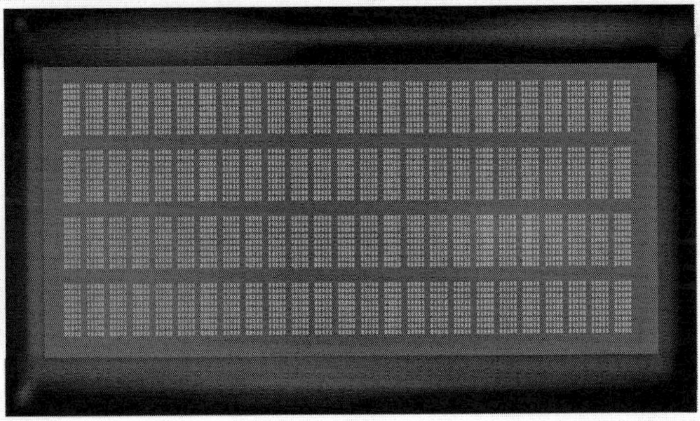

Figure 6.2: Infrared heater.

Source: https://pixabay.com/vectors/heating-infrared-ir-radiation-146210/

In comparison to conventional heating equipment, the equipment used in the infrared heating process has lower operating and maintenance costs and this process is much safer and cleaner. In infrared heating process, a desirable level of heating can be achieved at the surface as well at the core of the body. This technology is new in application in the industries, as it has always been restricted to the experimental processes.

There are a radiator and a reflector in the infrared heating component. The radiator radiates in all directions and the reflector directs the heat radiation towards the targeted location. The range for the maximum energy flux of radiator is from 50 kW/m^2 (long wave) to 4010 kW/m^2 (ultra - short wave).

Radiofrequency heating

Radio frequency is a type of innovative techniques which is based on electromagnetic heating. Thus, it has the ability to provide food of high quality and that too economically.

Electro heating can be further divided into two types such as direct electro heating and indirect electro heating. Direct electro heating is when for heating up the food, the electric current is directly applied to the food. Indirect electro heating is when in order to heat up the food, the electric current is first converted into electromagnetic radiation which thereafter produces heat within the product.

Radio frequency heating techniques are more successful in the non – food industry than the food industry. Examples for non – food industry in which this technique is successfully used is paper, plastic and lumber. The consumer demands tastier, cheaper, healthy such as no – fat, chemical-free and much safer product to consume. Because of this, the application of radio frequency heating techniques in the food industry is now extending.

Radio frequency heating techniques are not preferred in some modern industries for some processes such as product cooking, baking, pasteurization, drying etc. The mechanism used in radio frequency heating techniques is that when the product is kept under a radio frequency environment, the molecules within the product reorient themselves (27 million times/s at 27 MHz) continuously. This reorientation is the result of the field applied around the product.

Because of the frictional interaction between the molecules, the reorientation of the molecules results in the bulk or volumetric heating within the product and not in the air or any such equipment around it. By using a combination of dipole heating and electric resistance heating, radio frequency heating can be achieved. The heating in these techniques is the result of the movement of the dissolved ions that are present in the food.

Like other electromagnetic heating techniques, the radio frequency with electromagnetic waves ranging from 30 MHz to 300 MHz has the additional advantage and that is of uniform heating in the foods and also, the high penetration depth which can be used in order to sterilize and pasteurize liquid products.

Another advantage of using radiofrequency heating technique is rapid heating, that allows faster line speeds and shorter line length for unit operations that are semi-continuous as well as continuous in the food industries.

Because of the advantages like rapid heating and uniform heat distribution, high penetration depth and reduced energy consumption, the techniques using radio frequency heating has been proved to be the promising techniques for food applications. The applications of radiofrequency can be for softening frozen meat and in meat processing. The application of radiofrequency heating in pasteurization and sterilization of food is much restricted.

Radiofrequency heating techniques can also be used in disinfestation of food products. The radiofrequency heating is the best alternative for the chemicals that are banned such as Methyl Bromide and Phosphine that were previously used for disinfestation of post-harvest pests in food products such as semolina, spices, pet food, corns, etc.

Figure 6.3: Spices used in cooking.

Source: https://www.flickr.com/photos/30478819@N08/45395875502

Radiofrequency heating has evolved as the solution for improving shelf life. When a product contains average moisture, it is not necessary that the moisture content is uniformly distributed within the product.

For example, after the baking of biscuits is done and when the product contains 1 % of residual moisture, it is ready for being packed. Mostly, there comes a case when the center of the biscuits contains 3 % to 4 % of moisture. When such packed products are brought to pass through the distribution system, it only lasts for either one or two months before its consumption.

This condition can be prevented by passing the product through the radio frequency heating in order to increase its shelf life by evenly distributing the moisture and reduce it.

Ohmic heating

Ohmic heating is also referred to as joule heating, electrical resistance heating, electro heating, electro conductive heating or electrical resistance heating. In ohmic heating, the electric current is made to pass through the food or other material which is required to heat.

Ohmic heating is different when compared to other electrical methods such as microwave heating or radiofrequency heating. The difference is of the frequency range of the electric field and also, electrodes are present in the ohmic heating and not in other heating processes.

Ohmic heating consists of electrodes which are considered to be the source of power and a chamber to hold the food sample. The heaters in ohmic heating equipment can run on two modes, it can be either static mode or continuous mode.

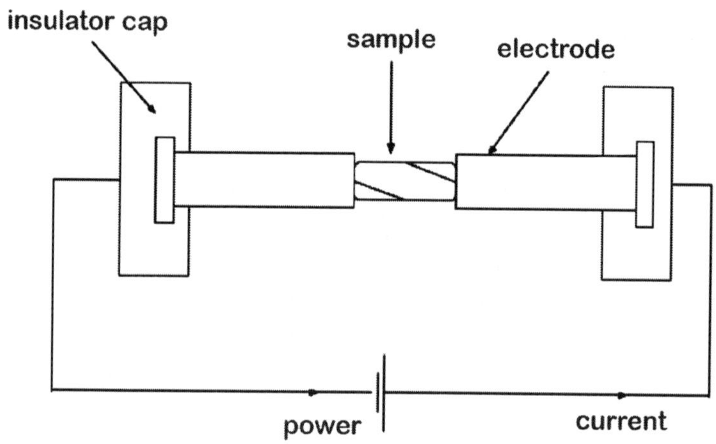

Figure 6.4: Ohmic Heating Process for Food Processing.

Source: https://commons.wikimedia.org/wiki/File:Ohmic_Heating_Process_for_Food_Processing.png

A typical static or batch heater includes a horizontal cylinder in which one electrode is placed at each of the extreme ends. The continuous ohmic heater has more flexibility and the design of the ohmic heaters includes distance between the electrodes, electrode configuration, heater geometry, electrolysis, requirement of power, the frequency used for AC current, applied voltage, current density, velocity profile and product velocity.

Ohmic heating techniques can be applied to a number of varieties of foods such as liquid, semi-solids or semi-liquids kind of foods. Ohmic heating is used in order to pasteurize the liquid egg products in US and for processing the fruits in the UK and Japan.

The techniques including ohmic heating are widely used in a variety of commodities such as vegetables and fruits, meats, soups, juices, seafood and some pasta dishes. In order to produce high-quality low acid as well as shelf-stable food, ohmic heating is used to sterilize foods such as ready to eat kind of meals and high – acid foods such as tomato – based sauces.

Ohmic heating requires the main parameter for heating purpose and that is electrical conductivity. The moisture content and the concentration of ions present in the products are the factors on which the heating effect depends.

For a number of foodstuffs, the electrical conductivity varies and also, it depends upon the temperature and voltage gradient used. The electrical conductivity, for purely liquid foods, increases undeviatingly with respect to temperature but overall falls with the increase in the concentration of the pulp.

The electrical conductivity of the solid foods increases linearly with respect to temperature at low voltage gradients. It can also vary in different directions within the solid. The electrical conductivity in the foods can be handled by changing the concentration of ions present in the food.

Ohmic heating techniques provide rapid as well as uniform heating. The surface does not become hot while ohmic heating which prevents thermal damage to the food product. The food which is most suitable for ohmic heating is the food which contains water and ionic salts in noticeable amount.

Sous-vide processing

Sous vide is a French word which means 'under vacuum'. Therefore, sous-vide processing is a process of cooking food which is sealed packed in pouches in low-temperature water bath. Unlike other cooking methods, this processing provides texture and doneness to the food.

The sous vide processing cooks the food at a temperature will which it will be served. While comparing sous vide cooking with other processes, it evenly cooks the food at a perfect temperature in order to provide texture to the food.

Figure 6.5: Saus Vide Cooking.

Source: https://commons.wikimedia.org/wiki/File:Sous_Vide_Cooking.jpg

There are a number of benefits of using sous-vide cooking procedure:

- A small amount of additional fat is required because the food is cooked at a low temperature.
- By using sous-vide cooking process, there is a lesser loss of nutritional contents from the food as compared to traditionally prepared food due to reduced oxygen in the environment.
- For cooking tough meats, the sousvide method is very useful. It helps in softening the meat by cooking it at low temperature.
- Using this sous-vide method for cooking reduces the sodium content in the food. While cooking at low temperature, the flavor is automatically enhanced and thus it required less salt and spices.
- This process maintains the smooth texture of the food by sealing the product in vacuum pouches that will dense the food.
- When meats are overcooked, it dries up. Sous-vide process prevents this condition thus, cooking meat at low cooking temperatures.
- The process gives consistent and desirable results as the cooking time and cooking temperature are carefully monitored and controlled.

Aseptic processing

A process that involves pumping of pumpable food material by using a set of heat exchangers where the product is heated up rapidly under pressure greater than or equal to 130°C is called as aseptic processing. This is a continuous thermal process which is used in order to produce shelf-stable foods.

After the product is heated, it is made to pass through a holding tube. In the holding tube, the temperature of the product mixture is equilibrated or balanced and thus held constant for a short time period considering the type of food and microbes present. Thereafter, it passes through a set of cooling heat exchangers in order to cool down the products. After this, the aseptically packaging of the sterilized cooled product is done in a reserialized package.

6.3.2. Non-Thermal Processing Technologies

High hydrostatic pressure processing

A cold pasteurization technique in which those foods are processed which are previously sealed in flexible and water – resistant packaging and a high level of hydrostatic pressure is applied up to 600 MPa / 87, 000 psi for a short period of time is known as high-pressure pasteurization technique. The hydrostatic pressure is the pressure transmitted by water. The effect on the food is the same as the result of subjecting food to an ocean of 60 km depth.

In 1990, the first high pressure pasteurized product was developed and until 2000, the processing technique was successfully applied in most of the food industries.

High-pressure processing does not much affect the ingredient and it also maintains the freshness of the food by maintaining characteristics such as flavor and nutrients. It is considered as an alternative for traditional thermal and chemical treatments.

High hydrostatic pressure technique is minimal or non - thermal technology. It is also a key tool for the product of microbiologically safer and shelf-stable fruits and vegetables. The amount of pressure applied and the time for which it is applied affects the inhibited microorganisms and enzymes that can cause deterioration.

If the temperature is not increased above the pasteurization temperature, safe fruits and vegetable products can be produced without any change in the flavor. In order to increase the shelf life of products, a combination of obstacles such as refrigeration temperatures, anti – browning agents and low pH are used, then this can cause a delay in the activity of remaining enzymes in the high hydrostatic pressure processed fruits and vegetable products.

High hydrostatic pressure processing is proved to be a relevant minimal or non - thermal energy technology which provides a large variety of not so more processed fruits and vegetable products than the demands of the consumers.

For processing and preservation of foods along with regaining their nutritional qualities, high hydrostatic pressure processing is used at a wide range of pressures and temperatures.

Irradiation Processing

In order to disinfect the food, the ionizing radiations are used. In the process of irradiation, the molecules of product absorb energy in order to form highly unstable and reactive ions or free radicals. These free radicals cause the breaking of a small percentage of chemical bonds.

Electrons that contains high energy breaks down the molecules thus forming highly reactive species like radicals. These radicals disturb other cellular macromolecules such as proteins, DNA or cell membranes. During the active replication, the cell genetic material becomes sensitive and responsive to radical damage.

Irradiation process contains equipment that has a high – energy isotope source so that it produces γ-rays. The most often commercially used source for obtaining γ-rays is Cobalt – 60 (60Co). There are also some machines which produce electrons with high – energy and X – rays but these machines are less commonly used.

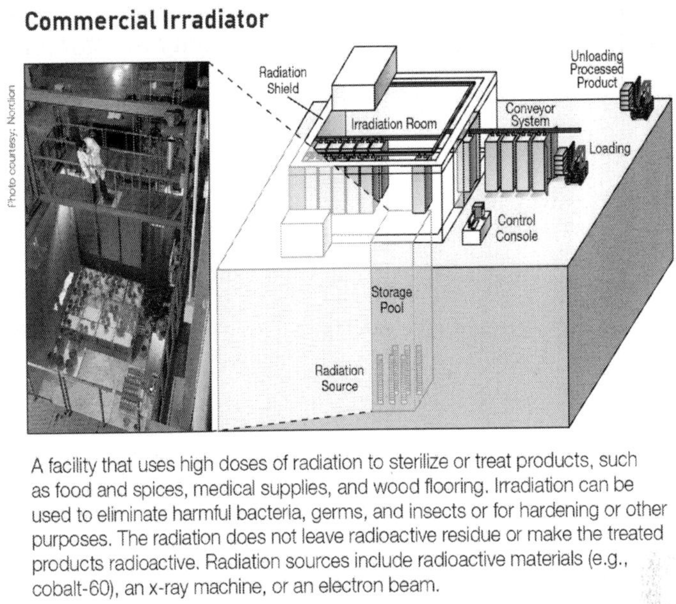

Figure 6.6: Commercial Irradiator.

Source: https://www.flickr.com/photos/nrcgov/36801710915

Example of machine sources is electron accelerators. The electron accelerator consists of a heated cathode and an evacuated tube. A heated cathode is used to supply electrons and in the evacuated tube, the electrons are accelerated because of the high – voltage electrostatic field.

Transmission of energies takes place by the sources of ionizing radiation, by evicting atomic electrons. These atomic electrons ionize other atoms by using a self – amplifying cascade of collisions. The source thus used depends upon several factors such as the absorbed dose, thickness and density of the food and the overall details of the process.

In comparison with the γ-rays and X – rays, the high energy electrons have lower penetration power and so it is used for only thin foods. Thick concrete walls and lead shields are used for preventing the leakage of the radiation outside the processing area. While performing irradiation, strict procedures are followed and it is carefully monitored to make sure the safety of the personnel.

In order to extend the shelf – life by about 2 -3 times, some types of fresh products such as strawberries and tomatoes are processed under irradiation. They are stored at 10 °C and so slow down the ripening process.

By obstructing the production of hormones and interfering the cell division and growth processes, irradiation controls the ripening and maturation of the fresh products. One more advantage of irradiation is, it can be used for disinfestation purposes. Irradiation at a dose of 3 – 30 kGy is recommended for the decontamination of herbs, dried spices, dry ingredients and vegetable seasonings.

Ultraviolet light Processing

The ultraviolet spectrum is divided into three regions having different wavelength range:

- UV – A, wavelength range: 320 to 400 nm.
- UV – B, wavelength range: 280 to 320 nm.
- UV – C, wavelength range: 200 to 280 nm.

When the DNA and RNA absorb ultraviolet rays, then the inactivation and cell death take place. Ultraviolet rays are used for surface treatments only because it has very low penetration depth.

The low vapor pressure mercury lamps produce the ultraviolet light. Special quartz glass is used in the making of ultraviolet lamps since it allows 70 to 90 % of ultraviolet rays to pass through it. In mercury vapor lamps, there is an inert gas carrier and a sealed glass tube filled with a small amount of the mercury. The creation of an electric arc causes the mercury to vaporize and ionize and so emitting the ultraviolet radiation. Altogether, these form the basic components of an apparatus that emits ultraviolet rays.

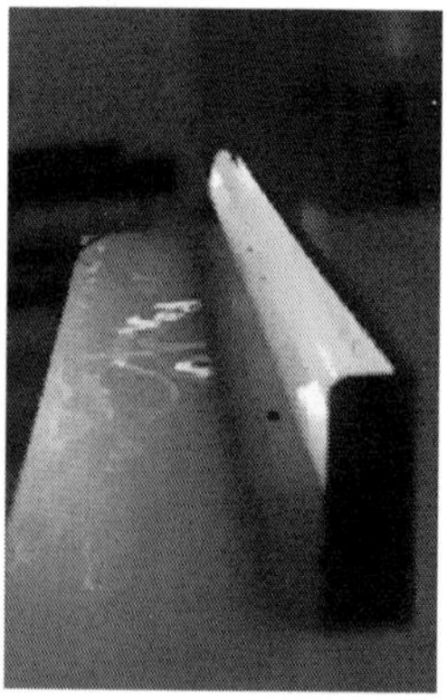

Figure 6.7: Ultraviolet light Lamp.

Source: https://commons.wikimedia.org/wiki/File:UV-lamp.jpg

The UV – C light technology is used for food products. Majorly, its use is restricted to liquid foods and water. When the UV – C light is made to fall into the liquid food materials, it showed that the UV – C light was able to travel a distance of only approximately 1 mm beneath the surface of the liquid food product and that too only the 90 % of the light energy.

The penetration effect of UV – C radiation is dependent on various factors such as the UV – C absorptivity of the product, the type of product, suspended matter in the liquid and the soluble solids in the liquid. The intensity of the penetration of the UV – C light will be low, if the amount of solid soluble in the liquid is larger as well as if the color of the medium is darker.

Ultraviolet light contains the ability to form superoxide radicals, hydrogen peroxide and lipid radicals. If the ultraviolet light treatment is extended beyond the limit, then it can produce peroxides which can affect the fat – soluble vitamins and pigments.

Furthermore, the superoxide radicals can encourage the cross-linking of carbohydrates, cross-linking of proteins, peroxidation of unsaturated fatty acids, protein fragmentation and loss of membrane fluidity function. This can cause harmful effects on the texture of certain foods.

Some vitamins such as vitamin A, B2 and C can also be degraded by using ultraviolet light by photodegradation. When the high – moisture foods are brought in contact with the ultraviolet light, then the water molecules present in the food absorb the UV photons can thus produce H^+ and OH^- radicals. These radicals can further interact with other food components. The UV radiation can also denaturalize the enzymes, aromatic amino acids and proteins that can lead to variation in the composition of the food material.

When certain products are exposed to UV light, it can cause a change in the flavor profile of the product. During the exposure, the radicals of oxygen thus formed affects the production of ozone which in turn leads to the development of food products without any flavor such as whipped cream, butter, milk, mayonnaise, sour cream and dried vegetable soups, etc. UV can cause the deterioration of product as well as several unwanted chemical reactions.

Pulsed light Processing

An innovative method which can be used for sterilizing and decontamination of the foods by using very high power and pulses of light emitted by inert gas flash lamps for a very short time period is known as pulsed light technology.

This process provides higher results in very short treatment times which makes it a very efficient technology to be used.

In comparison to ultraviolet processing, the pulse light processing is similar except that in this instead of rays or light, pulses are applied and that too at a high intensity. For the antiseptic action of the pulsed light, a number of mechanisms have been proposed. The ultraviolet part of the white light, which is emitted at high power, results in the inactivation of microorganisms by using a photothermal, photophysical and photochemical ways.

The combination of actions of light pulses causes harmful effects. Like because of the photochemical effect, the cell constituent such as nucleic acids and proteins is affected. Due to the photothermal effect, heat dissipation of light pulses, that penetrating the product, takes place which causes a rise in the transient temperature. The photophysical effect is caused by the constant disturbance produced from the high energy pulses which further causes morphological damage to the cells.

Use of pulse light technology can help in achieving effective microbial inactivation without causing any major harm to the properties of the products because it used light pulses for a very short duration but of high energy.

In order to obtain pulse light, capacitors release the power in the form of pulses of electricity to the lamps which are filled with inert gases. These pulses are of high voltage as well as high current. In order to get high conversion efficiency of electrical energy into light energy, inert gases such as krypton and xenon are used.

The current thus ionize the gas. This ionization results in a white light flash of broad-spectrum having a wavelength of 200 nm (UV) to 1 nm (infrared). The white light which is emitted contains approximately 45 % of visible light, 25 % of ultraviolet light and 30 % of infrared rays.

The pulse light is used in order to enhance the microbiological quality and helps in shelf life extension of foods. By storing the food such as pizza, bagels and cakes at an atmospheric temperature under pulsed light treatment, the shelf life of these foods can be extended up to 11 days.

Pulsed electric field Processing

A process in which a high voltage electrical field (from 20 to 70 kV / cm) is applied across the food for a few microseconds, it is called pulsed electric field processing. There are several parameters that have to be considered in order to handle the microbiological safety of the processed foods. The parameters are treatment temperature, flow rate or treatment time, pulse width, pulse shape, electric field strength, pulse polarity and frequency.

Because of the electrical resistance heating effect, the temperature of the treated foods increases during the pulsed electric field treatment. This increase in temperature can affect the inactivation of microorganisms and other food properties.

Since the treatment is done at atmospheric temperatures, the vegetative bacteria are killed by the treatment, but the bacteria do not get inactivated. Components used in the pulsed electric field treatment are treatment chambers, fluid handling systems and pulse generators along with controlling and monitoring devices. The technology using pulse electric light can pasteurize a number of liquid foods such as soups, milk, juices and other beverages.

Ultrasound Processing

Ultrasonic waves use a frequency ranging from 0.1 to 20 MHz. They have the ability to influence the biochemical, microbiological and physical properties of food. The interaction of ultrasonic waves with the food depends upon the ultrasonic waves as well as product characteristics.

The process that uses ultrasonic waves can be considered as both either thermal or non – thermal treatment with respect to the mechanism of action. During the process of ultrasonication, the creation of regions of compression that is positive pressure and rarefaction that is negative pressure takes place.

Formation of bubbles occurs in the liquid during the region of rarefaction and they break down during the region of compression. Thus, it creates micromechanical shocks or shear forces that can disturb the functional and structural components contained in the cells that will ultimately lead to the cell lysis.

When the cavitation bubbles collapse, generation of zones of highly localized extreme pressures (greater than 108 Pa) and temperatures (approximately 4000 K) occurs which disturbs the structure of cells and inactivate the microbes.

High power ultrasound processing or sonication is proved to be a good technology for liquid foods for inactivation of the spoilage microorganisms.

6.3.3 Other food Processing/Preservation Technologies

Fermentation

Desirable biochemical changes in foods, with reference to nutrition or digestion, are caused by the process of fermentation. The process of fermentation makes the food safe to consume and tastier by controlling microbial or enzymes. Some of the examples of fermented food are yogurt, beer, cheese, pickles, etc.

Figure 6.8: Fermentation of Beer.

Source: https://www.publicdomainpictures.net/en/view-image.php?image=148932&picture=beer-fermentation

Bacteria such as Lactobacillus, Lactococcus, Streptococcus, Bacillus and Pseudomonas are the representative vegetative bacteria that are used in fermentation. The yeast and fungi are also used for the process of fermentation. The food supports controlled growth of these microorganisms which in turn modifies the property of food such as flavor, color, taste, texture, etc. via secretion of enzymes.

Extrusion

The process of converting raw material into a product with the desired shape and form by forcing the material through a small opening by using pressure is known as extrusion. Examples of such food products are pasta, snacks, ready to eat cereals and textured vegetable protein. Extrusion offers some major advantages such as adaptability, process scale-up, high productivity, zero effluents, energy efficiency and low cost.

There is a tightly fitted screw in the extruder which rotates within a stationary barrel. In order to transform the raw food material into the final extruded product, the thermal and shear energies are applied to it within the extruder. The pre-ground and conditioned ingredients enters the barrel. In the barrel, they are moved, mixed and then heated by several screws and barrel configurations.

Figure 6.9: Extrusion Processing of Food Product.

Source: https://commons.wikimedia.org/wiki/File:StrasbourgChurros.jpg

The food may be exposed to a number of unit operations inside the extruder. These unit operations include fluid flow, mixing, heat transfer, melting and size reduction. The product comes out of the extruder through a die where it usually puffs (if the process is done at a temperature greater than 100°C and pressure higher than the atmospheric pressure) and thus changes occur in the texture because of the release of steam and normal forces.

Baking

Baking is a process in which the fully developed flour dough is processed by using dry heat and cooked into a variety of baked products such as cake, pastries, cookies, scones, bread, pies, pretzels, crackers, etc. the dough undergoes a number of stages before the process of baking.

From the process of yeast fermentation of available sugars, carbon dioxide gas is produced. During the process of baking, transfer of heat takes place from the source in the oven to the surface of the dough by convection. From the surface, the heat gets transferred by the process of conduction.

Figure 6.10: Baking of Bread.

Source: https://pixabay.com/photos/bread-baking-food-eating-3639071/

As the food is kept under heat, the heat transforms the batter or the dough into a baked food. This baked food contains a firm crust and a soft center. The heat during the process of baking makes water to vaporize into steam. In the presence of water, gelatinization of flour starch occurs.

The gluten which is the protein network firmly holds the structure. The carbon dioxide gas provides rise to the dough and it collapses when the baking is done at a temperature of 450°C approximately. There are a number of biochemical changes that take place in the batter or the dough because of the baking temperatures. These changes include denaturing eggs and gluten proteins, gelatinizing starch and dissolving of sugar crystals.

The surface of the dough loses water which is also caused by baking. It causes break down of sugars and proteins on the surface of the food that is baked. This all results in the formation of brown color of the food and the desired baked flavor.

Packaging

The packaging is a very important factor that plays a key role in food preservation techniques. There are many functions of packaging such as handling or ease in transportation, preservation, marketing and containment, etc.

The quality and properties of the food that is achieved by using various process are maintained with the help of packaging. The food product is protected from microbiological contaminants and other environmental factors by using packaging. The prevention of light-induced changes in the stored food products and a reduction in the loss of moisture is done with the help of packaging.

REFERENCES

1. Balasubramaniam, V., Park, S. and Lamsal, B. (n.d.). *Principles of Food Processing*. [ebook] p.2. Available at: https://pdfs.semanticscholar.org/16ae/0c71347f0d4fee8604ff6041348a0903c556.pdf [Accessed 10 Jun. 2019].
2. Dubey, D. (n.d.). *Proteomics & Genomics*. [ebook] p.1. Available at: https://nptel.ac.in/courses/102103017/pdf/lecture%201.pdf [Accessed 10 Jun. 2019].
3. Guerrero-Beltrán, J., Barbosa-Cánovas, G. and Swanson, B. (2006). High Hydrostatic Pressure Processing of Fruit and Vegetable Products. *Food Reviews International*, [online] 21(4), pp.411-425. Available at: https://www.tandfonline.com/doi/abs/10.1080/87559120500224827 [Accessed 11 Jun. 2019].
4. Hiperbaric.com. (n.d.). *High Pressure Processing (HPP)Technology - Hiperbaric*. [online] Available at: https://www.hiperbaric.com/en/hpp [Accessed 10 Jun. 2019].
5. Jojo, S. and Mahendran, R. (2013). *RADIO FREQUENCY HEATING AND ITS APPLICATION IN FOOD PROCESSING: A REVIEW*. [ebook] p.1. Available at: https://pdfs.semanticscholar.org/a77b/656ab8a5926263a59e6ff9247d3e30bad091.pdf [Accessed 10 Jun. 2019].
6. Kumar, J., Ramanathan, M. and Ranganathan, T. (2014). *Ohmic Heating Technology in Food Processing*. [ebook] p.1. Available at: https://www.ijert.org/research/ohmic-heating-technology-in-food-processing-a-review-IJERTV3IS20660.pdf [Accessed 10 Jun. 2019].
7. Kumari, J., Chandrasekhar, K., Dileep, A. and Lebonah, D. (2014). *A Short Review on Proteomics and its Applications*. [ebook] p.2. Available at: https://www.scipress.com/ILNS.17.77.pdf [Accessed 10 Jun. 2019].
8. Neetoo, H. and Chen, H. (n.d.). *Alternative Food Processing Technologies*. [ebook] p.3. Available at: https://pdfs.semanticscholar.org/0b75/8f1ad44a255a445dcc24f24efc3c57f605b1.pdf [Accessed 10 Jun. 2019].
9. Piyasena, P., Dussault, C., Koutchma, T., Ramaswamy, H. and Awuah, G. (2007). Radio Frequency Heating of Foods: Principles, Applications and Related Properties—A Review. *Critical Reviews in Food Science and Nutrition*, [online] 43(6), pp.587-606. Available at: https://www.ncbi.nlm.nih.gov/pubmed/14669879 [Accessed 11 Jun. 2019].

10. Proteomis.com. (n.d.). *Functional proteomics | Proteomis*. [online] Available at: https://www.proteomis.com/en/functional-proteomics [Accessed 10 Jun. 2019].
11. Sous Vide Cooking. (n.d.). [ebook] p.3. Available at: https://cedarlaneculinary.ca/Content/pdf/PerfectforSousVideCook.pdf [Accessed 10 Jun. 2019].
12. The Functional & Chemical Proteomics Handbook. (2016). [ebook] p.1. Available at: http://www.biotechsupportgroup.com/v/vspfiles/templates/257/pdf/FunctionalProteomicsHandbookNRicher032316.pdf [Accessed 10 Jun. 2019].
13. Wate, D. (2015). *Radio frequency heating*. [online] Biscuit people. Available at: https://www.biscuitpeople.com/magazine/post/radio-frequency-heating [Accessed 11 Jun. 2019].

7

Proteomics and Food Safety: Food Pathogens and Allergens

CONTENTS

Food safety is a leading global concern. Various techniques have been developed to assess food safety. Two factors that pose a risk for food safety are presence of microorganisms and allergens. Proteomics is science concerned with study of proteins that help identify the presence of these factors. This chapter discusses about application of proteomes for identification of food pathogens and protein allergens. The chapter discuses about microorganisms that cause food spoilage in various products. The chapter then discusses how proteomic approaches helps in identification of pathogens. The chapter then throws light on how allergens can pose a significant threat to allergic consumers. The chapter then describes in detail the application of proteomics in food allergen-omics covering the important techniques and approached used.

7.1 INTRODUCTION

Food safety is a serious global concern throughout the world. It not only results in losses but also poses a significant danger to consumer health. A reliable assay is required to evaluate and maintain food characteristics. The importance of a fast and reliable analytical method for monitoring the safety of food through the food chain cannot be stressed enough.

One of the main factors of food spoilage is pathogens and allergens. There are certain bacteria and some fungi which produce toxins and cause illness in people who consume them. New risks are arising because these microbial species are also adapting to various techniques that are used in hindering their growth. They are further aided by a change in food processing techniques, environmental factors and import and export of foodstuff.

As a result, while earlier the safety concern was mainly regarding food from animal origin, now even food of plant origin is under consideration. Proteomics offers considerable opportunities for analysis of pathogens. It gives an insight into pathogen activities during infection, an outbreak of disease and while recovering from it.

There are various factors that determine microbial activity in any food such as storage condition, inherent characteristics of the food, handling methods, etc. The natural microflora that are present in the food also determines whether the pathogen will spread. Proteomic techniques enable identification of cellular biomarkers to understand the adaptive capacity of pathogens in various conditions. For example, it was recently identified that pathogens form biofilm in order to survive.

Proteomics can present a powerful tool that can help in monitoring food safety due to its quantitative and qualitative features. It is an important branch of science which helps is assessing food safety and quality. In the above example, proteomics can help in identifying biomarkers typical for biofilm growing microbial cells.

Another safety concern with respect to food is a food allergy. It is a relevant health problem which may have many effects ranging from mild rashes to severe life-threatening reactions. Advanced techniques of proteomics can identify the causative food allergens from the food sources. It helps in assessing the physicochemical properties of food allergens.

This chapter explains the application of proteomics for ensuring food safety with respect to microbial contamination and the presence of allergens.

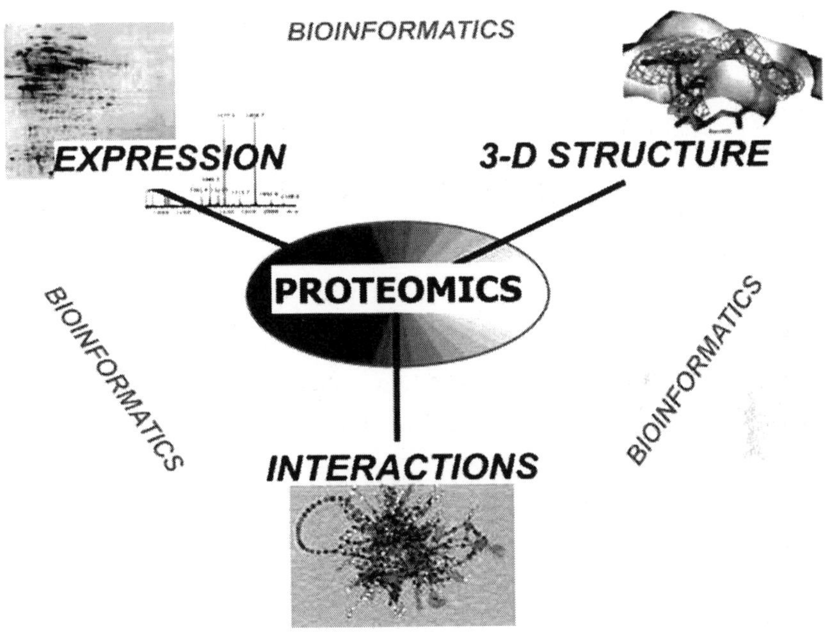

Figure 7.1: Representation of the Mechanism of Proteomics.

Source: https://upload.wikimedia.org/wikipedia/commons/1/14/Proteomics.jpg

7.2 MICROORGANISMS CAUSING FOOD SPOILAGE

One of the causes of food spoilage is pathogenic contamination. The number and the types of microorganisms affecting food depend on the following factors:

- The environment in which the food is kept
- The intrinsic factors and the microbiological quality of the food in its raw state
- The safety measures taken when handling and processing food
- Other external factors like packing, handling and storage that can prevent microbial growth

7.2.1 Meat Products

The microbes that are present in the fresh meat are:

- Gram-negative aerobic rods such as Pseudomonas, Acinetobacter and Moraxella
- Bacillus and clostridia

The subsurface portions of meat products generally do not contain pathogens, however, certain organs like the lymph nodes may have a significant presence of microbes. During food processing, the microbes may spread throughout the product. Hence fresh is one the most perishable food.

The meat spoilage does not cause a breakdown of the primary protein structure. The bacteria use free amino acids, glucose or other nitrogenous compounds and attain the population of 10^8 CFU/cm^2. This is when the meat spoils.

Ground meat has the same microflora as whole meat, but with higher loads. This is because it has a greater surface area over which the microbes can spread. When one piece of the meat is affected, grinding the entire meat together contaminates the entire lot. Besides, every processing step increases the risk of microbial contamination.

In case of vacuum packaged meats, some O_2 remains, which is consumed by the microorganisms for respiration. This results in higher levels of CO_2 and thus increases the shelf life. The microflora of vacuum packaged meats shifts from predominantly G- aerobes to G+ anaerobes and microaerophilic lactic acid bacteria (LAB) like *Lactobacillus*, *Carnobacterium* and *Leuconostoc*. Vacuum packaged meats have very less chance of getting contaminated except *S. aureus* and *Y. enterocolitica*

H_2S can be formed in vacuum packed fresh meat. H_2S forms Sulphur-myoglobin by combing with myoglobin in meats which have a pH higher than 6.0. This H_2S is produced by *Shewanella putrefaciens*, *Pseudomonas spp* and some *lactobacilli.*

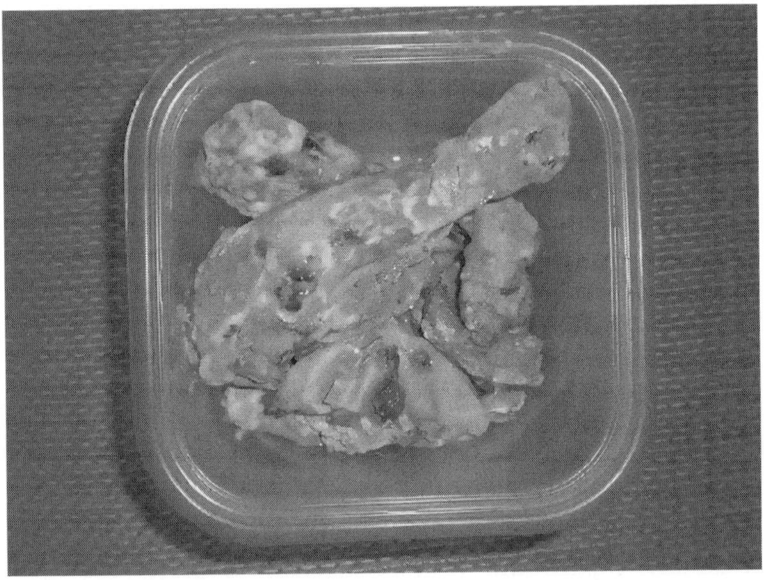

Figure 7.2: Representation of Meat spoiled by Microorganisms.

Source: https://commons.wikimedia.org/wiki/File:Moldy_chicken.JPG

7.2.2 Poultry

Food spoilage trends observed in poultry is similar to fresh meat. It is inhabited by similar microflora. In the advanced stage of spoilage, the skin of the poultry becomes fluoresce under UV because of the presence of fluorescent pseudomonads.

In the initial stages, the skin gets affected first than the body tissue. Hence in some cases, peeling the skin can help salvage the food.

7.2.3 Fish

Fish contains a high amount of nitrogen, but no carbohydrate. The microbial quality of fishes depends on the quality of water which they inhabit.

The microbial contamination in fish in majority of the cases is caused by un-sanitized processing.

The frozen fishes in general have less microbial flora than fresh ones. The fresh fishes have bacteria in gills, intestine and outer slime. Gill is the region which gets affected by bacteria first. Hence the best way to detect the presence of bacteria is to sniff this region. If it is spoiled, it can be identified by the odor emitted by *Pseudomonas* and *Acinetobacter-Moraxella* bacteria.

This bacterium then quickly move to the abdominal cavity. Spoilage of crustaceans and Mollusks occur in a similar way. In the case of mollusks, microflora depends on the water quality in which they are bred. Shellfish are filter feeders and usually carry all microbes that are present in their habitat. In case of fresh-water shellfish, *Pseudomonas* and *Acinetobacter-Moraxella* commonly occur.

7.2.4 Vegetables

Most vegetables have a pH of 6.0 which is conducive for bacterial growth. Vegetables contain molds, yeasts and bacteria. Almost 20 percent of the harvested fruits and vegetables are spoiled by microbes. Bacteria grow faster than fungi and are the leading cause of spoilage of vegetables.

The microflora of vegetables mainly comprises of lactic acid bacteria like leuconostocs, lactobacilli, streptococci, etc. *Staphylococci* may also be present though they generally do not spoil the vegetable. However, they may get transferred to other foods where the conditions are conducive. Bacterial soft rot is generally caused by *Erwinia carotovora* or *Pseudomonas spp*. In the case of low pH vegetables, bacterial growth does not occur. In these cases, molds are the spoiling agents. They enter through cracks or bruises.

Spores may get deposited by insects in these cracks. Botrytis cinereal is a type of mold which causes grey mole rot in many vegetables. There are various microbes which occur in the soil such as *C. botulinum*.

7.2.5 Fruits

Fruits have low pH around 1.8-5.6. Fruits are also rich in nutrients like vegetables, but their pH level does not support bacterial growth. Molds and yeasts play a greater role in fruit spoilage. Yeast grows faster than molds and often are the initiators of spoilage. Molds degrade polysaccharides and complete the spoilage process.

The common microbes are *Penicillium, Phytophora, Aspergillus* and *Geotrichum candidum* which causes blue rot, downy mildews, black rot and sour rot, respectively.

7.2.6 Dairy Products

Milk is a rich medium for microflora containing microorganisms that are present in the cow skin, udder and the utensil used. When milk is pasteurized and handled appropriately, it mainly carries G+ bacteria. Psychotropic pseudomonads are found in raw milk stored in bulk which reduces its shelf life.

In the pasteurization process, the majority of the G-, molds, yeast are removed while some G+ bacterial spores survive this heat. Another commonly found bacteria in raw milk is Psychotropic *Bacillus spp*. Previously, pasteurized milk was usually contaminated by LAB such as enterococci, lactococci, or lactobacilli. It reduced the pH level to 4.5 and was the cause of milk turning sour.

Today, the cause of spoilage of pasteurized milk are aerobic spore formers such as *Bacillus*. They release proteolytic enzymes which cause the milk to curdle. In rare cases, molds may appear on the surface, but milk is generally disposed of before mold contamination occurs.

Butter has high lipid content and low water activity (aw) which is more conducive for mold than bacterial growth. Some pseudomonads can also cause spoilage by producing organic acids, forexample *P. putrefaciens*.

In cottage cheese, bacteria, molds, and yeast cause spoilage. *Alcaligenes spp*. is the most commonly occurring bacteria which results in "slimy curd" appearance on the surface. *Penicillum, Mucor* and other fungi are also the causes of spoilage.

7.2.7 Eggs

Eggs have an inherent property like a hard shell and membranes like lysozyme, conalbumin which protect it. Egg white has a high pH (>9.0). These factors prevent the microbes from attacking the nutrient-rich yolk. The eggs are sterile when they are freshly laid but bacteria contaminate it very soon.

Eventually, these microbes enter the egg and initiate spoilage. The common spoiling agents are *Pseudomonads. Besides, Penicillium* and *Cladosporium* can also grow in the air sac and shorten their shelf life.

7.2.8 Cereal and Bakery Goods

These foods have low aw. Hence if it is stored suitably, it hinders microbial activity, except mold. *Rhizopus stolonifer* are other related species are found in bread, cereals and other baked foods. Refrigerated frozen dough products have more water content and can be contaminated by lactic acid bacteria.

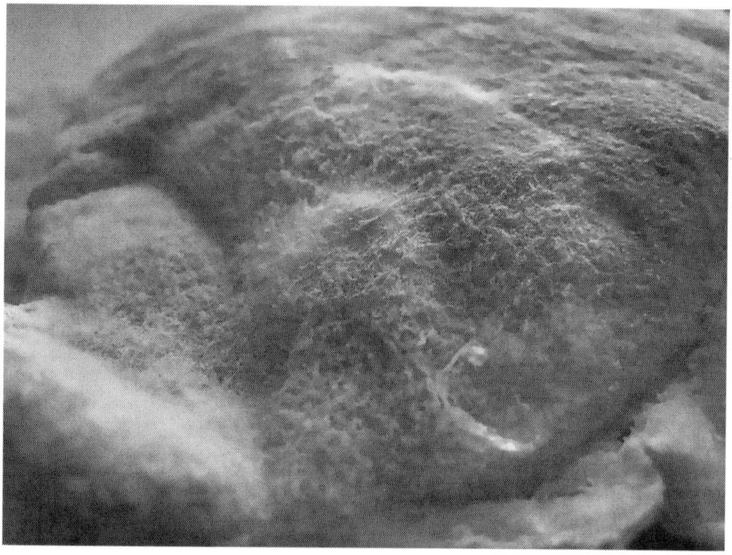

Figure 7.3: Representation of Moldy Bread.

Source: https://commons.wikimedia.org/wiki/File:Moldy_old_bread.JPG

7.2.9 Fermented Foods and Beverages

These food products have low pH which inhibits the growth of pathogens; however, food spoilage can occur. Beer and wine have a pH level of 4-5 and can be contaminated by bacteria and yeast.

The common bacteria found in beer are lactobacilli and *Pediococcus* spp., and (under aerobic conditions) acetic acid bacteria like *Acetobacter* and *Gluconobacter spp* and *Megasphaera cerevisiae* are present. Packaged beer is spoiled by yeast-like *Saccharomyces diastaticus*.

In the case of wine, *Candida valida* is the cause of spoilage in most cases. Lactic acid bacteria can also spoil wine by converting malic acid into lactic acid.

Fermented vegetables are spoiled by yeasts, lactic acid bacteria and mold. These include vegetables like sauerkraut, kimchi and pickles. Acidic foods like salad dressings can also be spoiled this way.

7.3 APPLICATION OF PROTEOMICS FOR IDENTIFICATION OF FOODBORNE PATHOGENS

There are many food products which have a short life which requires a fast and economical technique for detecting pathogens. The traditional methods for detecting the presence of pathogens use multiple selective culture media. Moreover, they also may take up to a week for confirmed results.

Proteomics assisted techniques can help in detection of the pathogen in food matrices. It is a fast method and in some cases species-level identification is possible within 18 hours using MALDI-TOF MS analysis. Various studies have been conducted whose results will assist in developing advanced diagnostic tests that can identify foodborne pathogens.

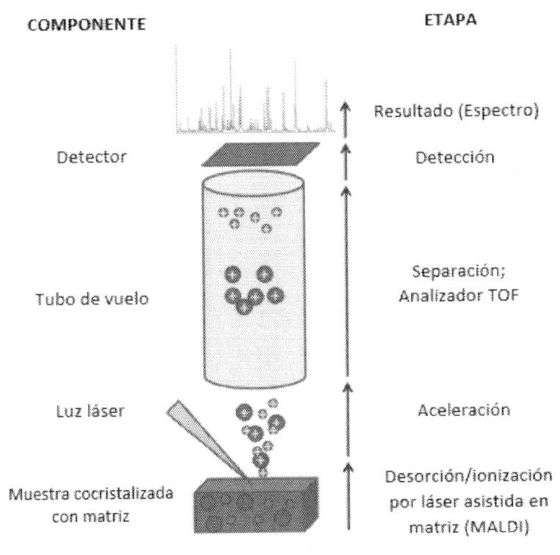

Figure 7.4: Representation of MALDI-TOF MS Technique.

Source: https://commons.wikimedia.org/wiki/File:MALDI-TOF_MS.png

7.3.1 Bacteria

Bacteria in food help in maturation and ripening, however in they could also pose a serious risk tofood safety. Proteomics plays a critical role in understanding how bacteria infect the food, develop antibiotic resistance and form biofilms to protect themselves. Proteomics has advanced significantly and provides many tools for this purpose. They help in detecting foodborne pathogens which can prevent spread of serious illness and other health hazards.

E. coli is among the leading causes of foodborne diseases and is responsible for health hazards and outbreaks. In order to address this issue a direct method detection of enterohemorrhagic *Escherichia coli* O157:H7 was developed by researchers. This method includes an enrichment step which involves magnetic beads that are coated with antibodies. The antibodies used are specific to antigens of serotype O157.

Post this enrichment step researchers used MALDI MS analysis and were able to detect 2 x 10(6) cells/ml. Recently, in another study, MALDI-TOF-TOF and tandem mass spectrometry were used for characterization of various species of Shiga toxin-producing Escherichia coli. This technique used a top-down approach.

Staphylococcus aureus is another foodborne pathogen and affects humans and animals alike. Mastitis could be found in milk and its products and is common in the dairy cow. The presence of subclinical mastitis is a pressing concern globally for the safety of milk. This bacterium affects the quality of milk and makes it unsuitable for deriving other products.

If the somatic cell count is high in milk, it indicates the presence of mastitis. The somatic cells are associated with a different amount of proteome and comprise of key components like lactose and fatty acids. Proteomics has played an important role in ensuring milk safety. It has provided many tools that can detect subclinical mastitis in milk and serum of infected cows.

In a recent study, the researchers found several potential biomarkers in milk and serum from affected cows. They are good candidates for the identification of subclinical mastitis. The techniques used for these studies were 2D electrophoresis and shotgun MS analysis. The biomarkers that can be used for detecting subclinical function, comprised of many proteins like serotransferrin, fibrinogen b-chain, and antimicrobial polypeptide (cathelicidin).

The study of MFGM was crucial for the development of this field. The MFGM technique used for bovine analysis demonstrated the differential protein expression that was found in cases of *Staphylococcus aureus* infection. *S. aureus* is a major pathogenic bacterium and is responsible for over 185,000 foodborne diseases every year in the US alone. It causes may gastrointestinal disorders like vomiting, nausea and diarrhea. The effects are visible in a few hours of ingestion of even a few milligrams.

Callahan and his co-workers created a technique based on mass spectrometry which can identify a protein toxin called staphylococcal enterotoxin B. This is a highly effective method that is based on detection of tryptic fragments using tandem mass spectrometry. It has been found that staphylococcal enterotoxin B can occur up to 5 parts per billion in foods that are soluble in water.

Another significant development in the absolute quantification of S. aureus toxins was a technique that combined immunocapture and Protein Standard Absolute Quantification (PSAQ) methods. It has huge potential to counteract huge outbreaks. *Listeria monocytogenes* is another widely present foodborne pathogen that causes serious illness and even fatality in humans. Traditionally, culture methods were used to detect its presence which took four to five days.

Researchers developed a sensitive and simple method that was able to detect *Listeria monocytogenes*. This was possible by the use of selective enrichment broth. The use of MALDI-TOF MS through this approach was able to detect up to 1 colony-forming unit of *L. monocytogenes* per ml. Moreover, the time taken was also reduced drastically and it took only 30 hours.

MALDI-TOF MS provides a powerful technique for pathogens detection in foods. This technique is based on the profiling of the whole bacterial proteome providing a fingerprint specific of the analyzed microorganisms in that specific time and physiological condition. This method obtains the fingerprint of the specific microbes analyzed. It helps in describing the features of the organism and characterize its strains and serovar. The advancement in the field of proteomics can help in detection of foodborne pathogens which cause serious illness.

Various experiments were conducted that have proved that MALDI-TOF MS technique is a suitable technique for detecting the presence of pathogens. It facilitates cost-effective and fast identification of various bacteria fin samples of food. It can also be used for testing human blood and serum.

In 2013, researchers were successful in identifying and classifying lactic acid bacteria that occur in fermented food. This method has been successful for identifying bacteria which is responsible for spoiling milk and pork. The advantage of this approach is that it takes less time and cost. The main challenge is in reproducibility.

The importance of proteomics is felt not only at a single species level, it is also important for studying bacterial communities. Their application for studying fermented food is especially important to gain insight into the dynamics of bacterial growth. It also helps understand the interaction of thousands of bacteria in the same habitat.

There is no doubt that mass spectrometric analysis holds huge potential in a complex environment. However, at present, the research of microbes using proteomics is limited to acid mine drainage, activated sludge, biofilms, soil, freshwater and seawater microbial communities, and human gut microbiota. The complex microbial systems in food must be researched much deeper using proteomics.

One pre-metaproteomic study investigated the release of bacterial protein inside Emmental cheese. The technique that was used was a combination of 2DE, protein fractionation and mass spectrometry. The cheese itself contains many proteins, which hindered the process. There were other technical limitations too. The researchers detected the presence of some bacterial species and several metabolic processes that caused the ripening of cheese.

The researchers made another attempt and used ITRAQ based approach this time. They succeeded in identifying and quantifying almost thirty proteins that were found in Swiss cheese. These proteins were both bacterial and bovine. In another study, the researchers investigated the metaproteome of fermented maize dough.

The method they used was a 1DE separation coupled to an LC-MS/MS of gel tryptic extracts on an LTQ Orbitrap Velos. They were able to identify hundreds and hundreds of proteins that belonged to plants, bacteria and fungi. Many species were identified that dominated this microbial ecosystem. The bacterial species were *Lactobacillus* and *Acetobacter* and the fungal species was *Aspergillus*.

At present, in order to gain more knowledge about the microbial community using proteomics, it is important to couple proteomics with bioinformatic strategies. Complete and well-annotated proteomic and genomic sequence is a major limitation for the researchers. This is despite the fact that NGS technology has made significant progress and can sequence all microorganisms which have enabled the creation of huge number of sequence database.

Additionally, NGS has high potential to perform advanced analysis of food composition and specifically of fermented and processed food. These are the food categories where microbial activity is the most. NGS method can be used to detect a group of one or more populations of an organism in processed foods. When it is combined with mass spectrometry it can identify the activities performed by the microbes in the sample tested. This information helps in gaining a better understanding of how processed foods undergo a transformation.

7.3.2 Fungi

Fungal contamination of food is a big problem in the food chain. For example, cereals are infected by mycotoxins. When the food is stored in warm and humid conditions, the molds can proliferate food and release mycotoxins in food. Mycotoxins are toxic secondary metabolites which are released in food by fungi. These fungi mainly belongto *Penicillium*, genera *Aspergillus* and *Fusarium.*

The types of mycotoxins found in food are ochratoxins, aflatoxins, zearalenone, trichothecenes, fumonisins, tremorgenic toxins and ergot alkaloids. They contaminate the crops, mainly cereals that are consumed by humans and animals. This way they penetrate the food chain. Mycotoxins have serious health consequences. They can result in cancer, renal failure, gastrointestinal diseases and mutagenicity. Some of the mycotoxins can also suppress the immune system and increase the risk of infections.

Proteomics has played a significant role in characterizing fungi metabolites and the proteins they secrete. This has helped in creating a complete database of the proteins secreted. The techniques used for detecting mycotoxin are immunochemical and chromatographic methods. At present, liquid chromatography along with tandem mass spectrometry is the most preferred method.

In the last decade, research in the field of proteomics has been focused to create a database and to investigate the various external factors present in the environment of the fungus can influence fungi stress response or proteome changes during mycotoxin producing conditions. Recently, various proteomic based researches have concentrated on how mycotoxins are able to control immunity in vitro, induce selective toxicity and modulate mitochondrial dynamics.

In the last 5 years, fungal proteomics research has concentrated on filamentous fungi. The research was aided by the availability of multiple genome sequences along with advancement in technology. These technologies included the innovation of next-generation nucleic acid sequencing and advanced proteomic technologies like tandem LC-MS. These advancements have made it possible to identify high-throughput protein and functional assignment which was beyond comprehension a decade earlier.

This intense activity has been driven due to various reasons. The need for gaining insight into the clinical consequences of opportunistic fungal infection was felt. This was more important in the case of immunocompromised patients. Besides, fungi are plant pathogenic in nature. Moreover, their enzymes have the biotechnological potential for producing biofuel.

As a result, proteomics has become a crucial part of system approaches aimed at understanding the richness and complexity of biochemical processes occurring in fungi. This is because it has the capacity to yield definitive information on protein identity, localization, posttranslational modification and the accuracy of *in silico* gene model prediction.

There was very limited knowledge that existed regarding fungal proteomes. Hence significant research was focused on developing techniques that could extract protein from fungi. This was used to establish basic data containing information about the types of proteins and their quantification in fungi. Efforts have been channelized at cataloging mycelial, organellar and secreted proteins of several types of fungi present in the world.

These techniques are useful for both individual protein identification as well as shotgun' proteomics. The former uses SDS-PAGE or 2D-PAGE fractionation. The latter analyses the total protein digest from created by fungi using tandem LC-MS. This generates the protein data sets that it contains.

The dynamic characteristics of fungal proteomes have also been studies. These studies investigate the impact of carbon sources, gene deletion and antifungal drugs at a proteomic level.

Till date, ascomycete proteomic researches have surpassed ones carried out using basidiomycetes. This is because of the preponderance of the former involved in causing animal disease, and because many ascomycetes are so-called 'cell factories' for protein secretion. A surprising fact of these studies is the occurrence of 'hypothetical proteins,' or 'predicted proteins'. These pepper all fungal data sets that were derived from research of the ascomycete, or more especially basidiomycete, Kingdoms.

When a protein is identified, it should no longer be classified as 'hypothetical' since its existence is proved. Instead, it should be classified as 'unknown function protein' (UFP). The identification of functions of the various 'unknown protein functions' is a major challenge that fungal proteomics face. Strategies are required for investigating these functions.

Proteomics will be an integral part of future research to gain more information about the nature of fungi and its biotechnological uses. The main challenge is assigning biological roles to both in silico annotated, and unannotated genes. The emerging online resources are associated with robust analytical strategies like quantitative proteomics and RNA sequencing for assessing modified gene expression. These resources aids in the functional categorization of the gene and proteins of fungi. They can aid in dealing with this challenge of assigning biological roles that proteomics currently faces.

Researchers have implied that the identification of fungal protein by protein mass spectrometry is like assembling pieces of a mega jigsaw puzzle. Only after all the pieces have been found and put in place that one can assess the complexity of fungal proteomes.

7.3.3 Other Pathogens

Besides the disease-causing bacteria and fungi, there are other microorganisms like protozoa and helminths that also result in food safety issues. The major pathogens in this category are *Giardia intestinalis*, *Toxoplasma gondii* and *Taenia spp.* These microbes spread easily and are transmitted through soil, water or contact with infected persons and animals.

There are increased incidences of their transmission and happens through mediums like undercooked meat, undercooked fish, crabs, raw vegetables and mollusks which get infected by excreta of humans and animals. Cryptosporidiosis is a serious infection in humans occurring in gastrointestinal organs that could represent a real burden for immunocompromised patients. This infection cannot be treated as no drug has been invented to prevent it from spreading.

Researchers tried to find how sporozoite transmission occurs by investigating LC-MS/MS coupled with a stable isotope N-terminal labeling of both inside and outside oocysts sporozoites.

7.4 FOOD ALLERGENS

In the past two decades, significant efforts have been directed to gain insight into allergenic proteins in food from plant and animal sources. This is important for understanding their physiochemical characteristics and their effect on the immune cells. Based on these extensive research, allergen databases were built which can be accessed for information when needed.

These studies revealed that only some of the protein families cause food allergies. In plants, the common allergy-causing protein families are 2S albumins, seed storage proteins identified from seeds like sunflower seeds, sesame seeds and nuts like peanut, walnut, hazelnut, etc. Other proteins are the non-specific lipid transfer protein (nsLTPs) which are found in fruits like apple, peach and tree nuts.

The seed proteins from the cupin superfamily, 7/8S globulins and 11S globulins that are found in seeds and tree nuts are also known allergens. Another allergy-causing protein family is the Bet v 1 which are present both in monocot and dicot. These proteins are important which helps the plant upregulate resistance to protect itself from pathogens and also cope with environmental stress.

In animals, tropomyosin and EF-hand proteins are common. Tropomyosins from crustacean, mollusks and fish parasites trigger allergies in many consumers. An EF-hand domain is shared by several different Ca2+-binding proteins and enables either Ca2+ dependent signaling and/or Ca2+ buffering or transport. Another major fish allergen is parvalbumin which occurs in many fishes. Milk also contains allergen protein, caseins which trigger symptoms in lactose-intolerant patients.

Immunoglobulin E (IgE)-mediated food allergens are a serious concern and are a major cause of allergic diseases in infants and small children. They affect somewhere between 1 to 10 percent of the global population. There are many foods that can trigger minor reactions, but some of them can evolve into severe conditions like anaphylaxis.

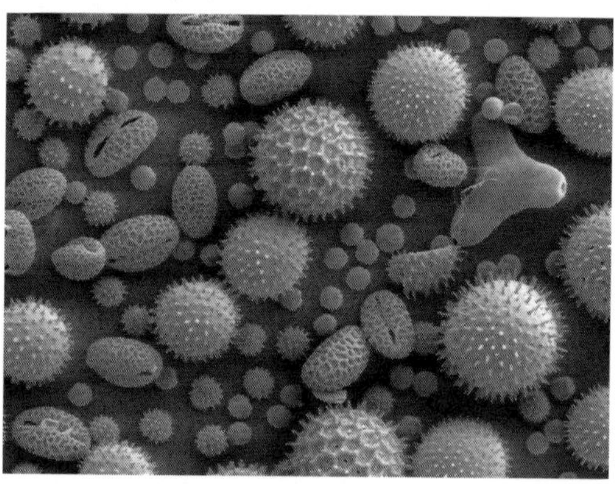

Figure 7.5: Representation of Food Allergens.

Source: https://upload.wikimedia.org/wikipedia/commons/a/a4/Misc_pollen.jpg

Anaphylaxis is triggered by food items like egg, milk, peanut, soy, fish etc. Egg, milk and soy generally affect children while nuts, seafood and fresh fruit allergies are found in adults. Though avoiding the allergy-causing food is the safest measure, with increased consumption of processed food, the effectiveness of this measure is reducing.

The food composition has become more complex, and food labels do not clearly indicate all the ingredients present. This leads to contamination of some allergens into the diet. Moreover, many production facilities produce multiple products. Hence although the product consumed may not have an allergen, some other food that was processed might which may trigger the symptoms.

Hence it is important to identify the allergens, state their characteristics and quantify their presence in all foods. Moreover, it is also important for their post-translational modifications and detects their molecular determinants that cause the allergy.

7.5 APPLICATION OF PROTEOMICS IN FOOD ALLERGENOMICS

Over the years, the techniques used to detect and quantify the allergens in food has evolved. Modern mass spectrometry technology and improved bioinformatic tools and genomic data banks have contributed significantly to the science of proteomics.

The scope of the application of proteomics in detecting protein allergens is huge. Critical information is derived from mass spectrometry-based proteomic analyses of novel FPA and the IgE–FPA epitope mapping. They help in the following ways:

- They help in understanding how immune reactions occur
- They help in creating novel and effective protocols that will help in the diagnosis and prognosis of protein allergens
- They help in analyzing the patient's clinical history
- They facilitate the safety assessment of food
- The enable post-translational modifications in food when the food is being processed so that its allergenic properties can be assessed.

7.5.1 Food Protein Allergen Detection by Commonly Used Methods

Historically, food protein allergens have been detected using antibody-based assays which may be complemented by protein sequencing techniques. The IgE from the allergic person is collected in these assays. The allergens are then identified through their direct binding.

IgE immunoblotting is a popular technique used for detection. This process uses gel electrophoresis, either one-dimensional or two-dimensional. The protein is then transferred to a membrane-like nitrocellulose or polyvinylidene difluoride. The serum samples are collected from allergic patients. The membranes are then incubated with those samples and the specific IgE binds the allergen.

This is an analytical technique which makes it possible to visualize IgE–FPA immune reactions, assigning the FPA of interest on a reference electrophoretic map. In the 1990s the Edmande gradation technique was integrated with this strategy which enables the determination of N-terminal amino acid sequence of allergens. This innovation helped in identifying the allergens in food through a sequence homology assignment.

Earlier, enzyme-linked immunosorbent assay (ELISA), was used to quantify protein allergens. Here the IgG from animal sources tethered the target protein allergens into microplate wells. This method allowed visualization, sequencing and quantification of protein allergens but had the following limitations:

- It was based on the complexity of the food matrix
- Mixing of contaminants in the sample from other food
- Variable specificity of the antibody preparations of the commercial ELISA kit

Moreover, animal and plant genome databases were not available which hindered the molecular studies for characterization of allergens along with the IgE epitope mapping.

7.5.2 Food Allergenomics

Techniques based on proteomic mass spectroscopy are useful in clear detection and quantification of even single protein allergens and their isoforms. They are effective even in the case of processed foods which would have undergone physical and chemical alterations. This has an impact on its protein stability and hence traditional methods cannot detect the presence of allergens in such foods.

The precursors of food allergenomics application integrated matrix-assisted laser desorption ionization–time of flight. (MALDI-TOF) mass spectrometry with the Edman degradation technique for a better characterization of the apple and peach protein allergens. These, methods removed the allergens from fruit protein by reverse-phase high-performance liquid chromatography (HPLC)and separated each fraction by 1-DE gel. The IgE immunodetection and immunoblot inhibition assays used the serum of the allergic patient to identify the protein allergens. There characteristics were then determined by N-terminal amino acid sequencing and linear MALDI-TOF mass spectrometry analysis.

With advancement in technology, high-resolution mass spectrometry techniques were invented. Bioinformatic tools and genomic data banks were also improved. This made food allergenomics an effective tool for the characterization and quantification of protein allergens.

The modern Allergenomics was first applied in 2000 for investigating protein allergens using mass spectrometry. In this study, the wheat proteins were separated using the 2-DE approach. Then, immunoblot analysis was used to characterized using these proteins. The serum of 10 allergic patients with Baker's asthma was investigated using mass spectrometry analysis.

7.5.3 Food Protein Allergen Characterization by Allergenomic Approach

Food samples often contain marginal levels of protein allergens. Hence to describe their attributes, the proteins must be concentrated. In most cases extraction and precipitation of protein is performed. This enables the removal of other irrelevant material and helps to concentrate the traces of protein. The most common pre-treatment methodology is the combinatorial peptide ligand library (CPLL) technique.

Till date, the identification and characterization of new protein allergens is done using the qualitative gel-based proteomic approach. This technique used the 2-DE approach to separate the protein extract. The next step is to stain it using a protein-dye.

The gel is then blotted against the serum sample being tested. The positive IgE-protein allergen spots are compared with the gel plate. Each of it is hydrolyzed separately using specific enzymes. Liquid chromatography systems help in separating the peptides that are formed. Subsequently, mass spectrometry or tandem mass spectrometry is used for analyzing these peptides.

Qualitative allergenomics has certain drawbacks. Only partial characterization of plant and animal genomes has been possible and gene functions have not been annotated completely. However, homology searching on the reference proteome databases does enable identification of the protein. Qualitative analysis is a critical technique for ensuring food safety.

In case of allergy to cow milk, some infants may ingest cow milk protein through their mother's milk. These infants are sensitive to maternal milk elimination. However, mother's milk was never found containing traces of cow milk proteins.

Recent research using proteomic approach could identify intact bovine alpha-S1- casein in mother's colostrum. This was found as the allergy trigger in case of children exclusively breastfed.

In another research, it was revealed that heating the protein allergens in prawn increased its antibody reactivity and to the derived fragments as well. Another study demonstrated peanut protein structures were altered when they were processed thermally. As a result, the majority of the immunochemical and targeted screening test prevented the accurate detection of protein allergens.

7.5.4 Food Protein Allergen Quantification by Allergenomic Approaches

There are many foods that trigger reactions in allergic patients. Hence, it is important to understand their levels in the food. There is an urgent need to quantify the level of allergens to ensure food safety. The methods for quantifying may be relative or absolute. Relative quantification identifies the differences in protein expression between two or more proteomes. Absolute quantification quantifies the number of proteomes in a given sample.

In the case of relative quantification, the methods are 2-DE based. It reveals modulations in protein expression by evaluating gels of various food. The 2D-differential gel electrophoresis was developed with the aim making all gels comparable. This method is effective in assessing changes in protein profiles from various samples by running it on a single 2D gel.

Another common technique used for protein allergen quantification uses the LC-MS/MS technique. Label based approaches are generally used for performing mass spectrometry bases approaches for relative quantification. On the other hand, absolute quantification is performed by label-free strategies. Recent novel technology is based on shotgun procedure and is referred as selected reaction monitoring. It does not use labels and adopts a targeted proteomic strategy.

The protein molecules are first extracted from the food matrix. They are then digested with the help of enzymes and then analyzed by LC-MS/MS approaches. It uses specific proteo-typic peptide for this. Selected reaction monitoring is now the benchmark process that is used for the purpose of absolute quantification of protein allergens. These results are traceable to the international system of units. It is a step towards a significant comparison of protein allergens data among laboratories.

In a study, relative quantitative changes of food allergens in cereals after the food processing was investigated. It was found that relative mass spectrometry-based allergenomics is universal. It is a reliable method and can detect quantitative changes in protein allergens while the food is processed.

The only recommendation in case of food allergy is avoidance from partaking those foods. Till date, no effective treatment for food allergy has been developed. However, allergen labeling at times lack certain information about the contents that can cause allergies.

Hence best practices for allergen risk assessments were developed by food manufacturers and regulators. This helped to examine training strategy for food manufactures and helped in identifying effective analytical techniques.

The mass spectrometry techniques are expected to enable the assessment of threshold limits of various allergens that are found in food. Recently, multiple reaction monitoring mass spectrometry was developed with the aim of quantifying allergens for every food item. Absolute quantification of proteins in food is important so that the risk of allergen levels an allergenic person is exposed to when consuming food can be assessed. It also helps in creating more refined and standardized assays that can improve allergenic risk management.

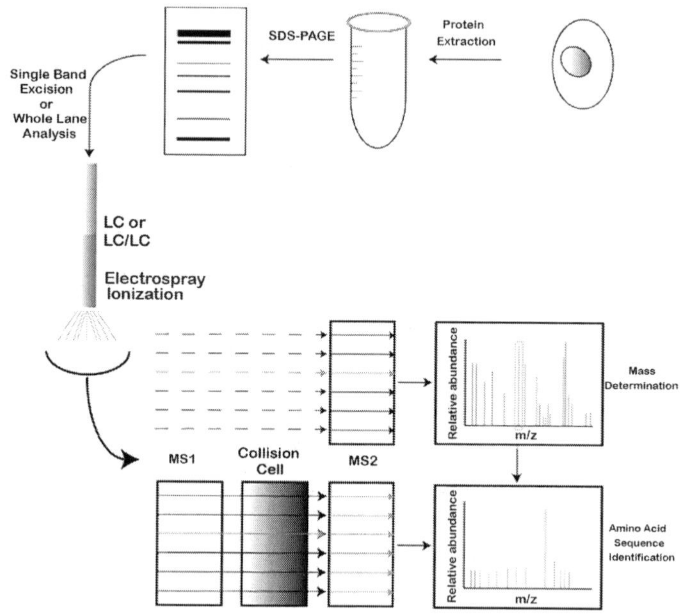

Figure 7.6: Schematic Depiction of LC-MS MS Procedure.

Source: https://upload.wikimedia.org/wikipedia/commons/0/03/Schematic_depiction_of_LC-MS_MS_procedure.jpg

7.5.5 Current Status and Future Perspectives for Quantification of Food Allergens

The allergens undergo a dramatic transformation when they are processed. Various foods have different allergic reactions, the food industry and regulatory bodies need dependable techniques that can do absolute quantification of allergens in complex food products which can safeguard the allergic consumers from foods that trigger their allergies.

The recently developed targeted proteomic techniques like mass spectrometry, multiple-reaction monitoring along with isotope-labeled internal standard referred to as AQUA peptides can provide absolute quantification of food allergens even at a level as low as 10 ppb.

In order to develop AQUA-MRM assays, certain pre and post-MS criteria have to be satisfied. The allergen levels may alter in animal and plant-based foods. Hence it is important to develop a high throughput multiplexed allergen quantification technique. This is important for developing an effective AQUA-MRM assay. Future strategies must be focused on making devices that will aid in better management of food allergy.

The rapid increase in food allergenicity makes it important to gain thorough knowledge of absolute allergen levels in food products. This information is vital to safeguard allergic consumers. In order to achieve this AQUA-MRM approach will play a significant role.

7.6 CONCLUSION

Food safety is an urgent issue that also has a global importance. It not only affects trade and economy, but also the health of humans and can have serious consequences. Prevention is better than cure and it is important to support manufacturers to produce food free from contaminants and allergens and is safe for consumption.

Proteomics can contribute significantly in detecting microbial contaminants and the presence of allergens. Proteomics can help develop techniques that can prevent foodborne pathogens from spreading by timely detection. The techniques can investigate the presence or absence of targeted proteins in raw and cooked food. This helps in detecting presence of pathogens. The proteomic techniques, specifically, MALDI-TOF MS analysis is playing a decisive role in evaluation of foodborne diseases.

The latest techniques of proteomics also facilitate the characterization of food allergens. This is useful for diagnostic application and far better than extract-based approaches. This can contribute to developing diagnosis methods that can be tailored to patient needs, help in better management of the patient by suggesting dietary changes, avoidance strategies.

It helps the food industry to establish quality standards and define threshold limits for protein allergens. The risk assessment strategies are also important for establishing food production and processing methods. Besides, novel foods are invented continuously, most of which undergo high processing. The also use food from sources that have not been tested so far. They should be tested using effective techniques since they potentially high allergenic risk.

The European Commission has funded COST Action “IMPARAS”. The aim is to find solutions to these issues. The Cost Action helped create a multidisciplinary approach with the help of proteomics researchers, nutritionists, immunologists and allergists. They help in tackling the challenges posed by allergenic risk assessment of foods consumed historically, and novel foods. Besides risk assessment of novel food, processing strategies and the impact it has on allergenicity of proteins present in food is also important.

REFERENCES

1. Anon, (2018). Detection of Foodborne Pathogens Using Proteomics and Metabolomics-Based Approaches. [online] Available at: https://www.ncbi.nlm.nih.gov/pubmed/30619201 [Accessed 11 Jun. 2019].
2. Anon, (n.d.). Microorganisms and Food Spoilage. [online] Available at: http://faculty.weber.edu/coberg/class/3853/3853%20mos%20and%20food%20spoilage%20notes.htm [Accessed 11 Jun. 2019].
3. Di Girolamo, F., Muraca, M., Mazzina, O., Lante, I. and Dahdah, L. (n.d.). Proteomic applications in food allergy: foodallergenomics. [ebook] Available at: https://www.academia.edu/13694651/Proteomic_applications_in_food_allergy_food_allergenomics [Accessed 11 Jun. 2019].
4. Doyle, S. (2011). Fungal proteomics: from identification to function. [online] Oxford Academic. Available at: https://academic.oup.com/femsle/article/321/1/1/634676 [Accessed 11 Jun. 2019].
5. Karin Hoffmann-Sommergruber, K. (2016). Proteomics and its impact on food allergy diagnosis. EuPA Open Proteomics. [online] Available at: http: https://www.researchgate.net/publication/301234808_Proteomics_and_its_impact_on_food_allergy_diagnosis [Accessed 11 Jun. 2019].
6. Piras, C., Roncada, P., M. Rodrigues, P., Bonizzi, L. and Soggiu, A. (2015). Proteomics in food: Quality, safety, microbes, and allergens. [online] Wiley Online Library. Available at: https://onlinelibrary.wiley.com/doi/abs/10.1002/pmic.201500369 [Accessed 11 Jun. 2019].

8

Proteomics and Systems Biology in Nutrition Research

CONTENTS

This chapter talks about Proteomics and Systems Biology and its relation with Nutrition Research with sub headings covering the topics like systems biology and its explanation, system biological research and food quality for personalized nutrition, nutritional systems biology modelling, the challenges involved in molecular nutrition research, as well as the complete relation between proteomics with nutritional research. This chapter laid down the basics of proteomics and protein structures with sufficient examples and structures. this chapter also discuss how biology can be relative and quite important in our day to day life in general way.

8.1 INTRODUCTION

In the span of last ten years, advances and innovations in genomics, proteomics, and metabolomics have harvested large-scale information that has spiked interest in global researches, with the main aim of understanding biological systems as a complete. Systems biology links computational modeling and researched biology to forecast and characterize the dynamic properties of biological systems, which are observed as complex signaling links.

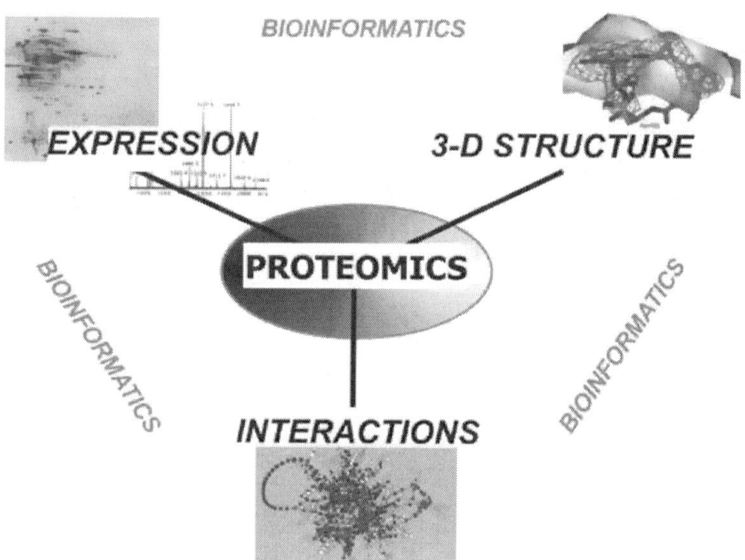

Figure 8.1: Proteomics basic structure.

Source: https://upload.wikimedia.org/wikipedia/commons/1/14/Proteomics.jpg

Whereas the systems research of disease-perturbed links holds good hope for detection of drug targets for therapy, equally the detected critical network nodes may be aimed through nutritional intervention in either a preventative or therapeutic method.

As such, in the same course of the nutritional sciences, it is predicted that systems research of normal and nutrient-perturbed signaling links in relation with knowledge of underlying genetic polymorphisms will spearhead to a future in which the health of a person will get better through forecasted and preventative nutrition.

Advances in MS have allowed for a linear upsurge in proteomic detections. Such approaches have allowed for the development of complex, multi-faceted information sets, and these will be the keystone of future systems biology analysis.

This chapter will explain systems biology, outline current proteomic methodologies, highlight successful applications of proteomics in nutrition research, and explain the challenges for future applications of systems biology methods in the nutritional sciences.

Research has allowed for advances in systems biology using sequencing and microarray technological innovations and more recently MS to allow analysis of different biological systems such as cells, tissues, biological fluids,

However, with the various sets of these datasets has emerged researcher's understanding that even extensive knowledge of the already in hand subsets of a system does not necessarily lead to understanding how a system functions or get executed. Such systems are shown to display "emergent" behavior with function arising from the totality of system interactions, which are often nonlinear and stochastic. Multiple cross-talk across the cells have been displayed by this approach with both feed-forward and feedback loops have been highlighted. The robustness of systems is allowed for analyzing the relationship between nutrition and health.

As stated in detail by researcher de Graaf that mathematical modeling has traditionally been applied in the nutritional sciences of which examples consists of metabolic flux analysis, compartmental models, and whole-body models of energy metabolism.

Although predictive network modeling in response to nutrient perturbations and or genetic polymorphisms has yet to be fully utilized in molecular nutrition research, the increasing application of the tools of genomics, proteomics, and metabolomics to nutrition-related research questions means that datasets of sufficient depth and complexity now exist for systems biology computational approaches.

8.2 SYSTEMS BIOLOGY

Within the duration of past years, the execution of systems biology in nutritional research has come up as a powerful apparatus to comprehend the mechanisms by which dietary components promote health and prevent disease as well as to detect the biologically active molecules being involved in such after effects.

Systems biology, by mixing several '-omics' disciplines which are mainly genomics/transcriptomics, proteomics and metabolomics, develops large data sets that upon computational linking deliver in silico predictive links that allow a deeper analysis of the personal response to a nutritional intervention and deliver a more worldwide comprehensive deep analysis of how diet may influence health and disease.

Many studies have projected that diet and in particular bioactive food, components play a pivotal role in helping to counteract environmental-related oxidative effects. Oxidative stress is being analyzed to be strongly implicated in aging and the pathophysiology of various diseases which includes neurodegenerative disease, cancers, metabolic disorders and cardiovascular diseases.

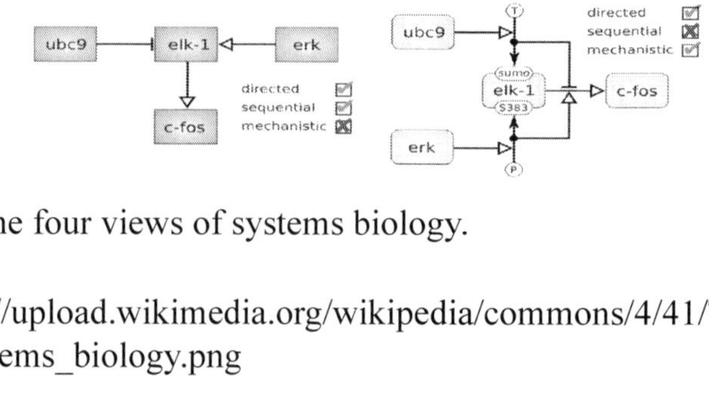

Figure 8.2: The four views of systems biology.

Source: https://upload.wikimedia.org/wikipedia/commons/4/41/The_four_views_of_systems_biology.png

In the upcoming sections, this chapter will provide insights into the role of systems biology in nutritional research and focus on transcriptomic, proteomic and metabolomics researches that have projected the ability of functional foods and their bioactive subsets to fight against oxidative damage and contribute to health benefits.

8.2.1 -OMICS Applications as a Foundation for Systems Biology in Nutritional Research

Nutritional research has its main attention on analyzing the link between diet and health in order to indorse natural ways of disease prevention. To this, nutritional end science deals with a broad range of analytical areas that has spread its wings from a better understanding of the molecular mechanisms of prevention and protection to the detection of the biologically active food components and the projection of their biological efficacy.

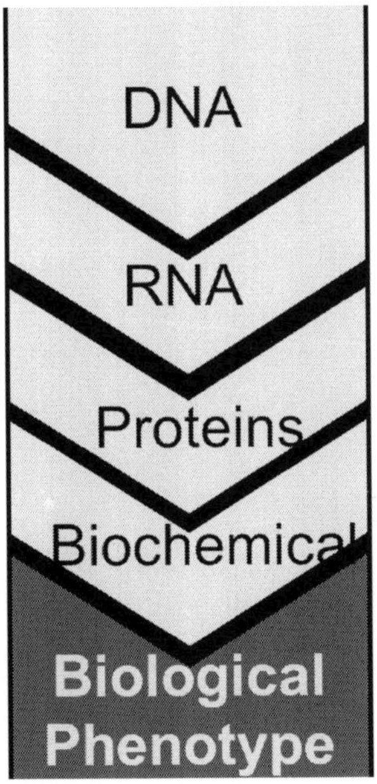

Figure 8.3: The central dogma of biology showing the flow of information from DNA to the phenotype.

Source: https://upload.wikimedia.org/wikipedia/commons/b/b8/The_central_dogma_of_biology_showing_the_flow_of_information_from_DNA_to_the_phenotype._Associated_with_each_stage_is_the_corresponding_systems_biology_tool_from_genomics_to_genomics_to_metabolomics.png

In the previous ten years, the implementation of systems biology has come up as a powerful apparatus to achieve these objectives. While a definitive and agreed-upon explanation of what consists of systems biology is still missing, systems biology can be explained as the mixture of various '-omics' technological innovations in different biological systems such as tissue, cells and/or biological fluids in order to develop large molecular information which, once linked in computational modelling approaches, deliver predictive links that is signaling links from the cell membrane to the nucleus that aim ultimately to analyze how the biological system under scrutiny behaves upon a perturbation for example dietary intervention.

In this scenario, the evolvement of three main '-omics' platforms which are transcriptomics, proteomics and metabolomics have made the research of how food impacts health status, to help in differentiating dietary responders from non-responders and to detect the nutritional bioactive components responsible for the health outcomes.

While transcriptomics delivers a comprehensive observation on all genes active at a given time point in a given prototype, the proteomic standard detects and analyses protein expression and characterizes post-translational modifications.

Although both disciplines may provide markers of efficacy and aim for intervention, proteomics does so in a more robust and efficacious way since variation induced in mRNA expression do not linearly interrelate with the changes observed at a protein level. Finally, metabolomics has its attention on the analysis of metabolites, their dynamics, composition and interactions and is considered to deliver the most functional information.

Without doubt, the linking in nutritional systems of gene and protein expression profiles with metabolic fingerprints makes a more extensive research of the individual answer to a nutritional intervention and delivers a more worldwide comprehensive understanding of how food components influence health status.

8.3 SYSTEM BIOLOGICAL RESEARCH ON FOOD QUALITY FOR PERSONALIZED NUTRITION AND HEALTH USING FOOD-OMICS TECHNIQUES

Safety once has been, and consistently remains without an uncertainty a focal point of consideration of consumers. By and by, with improvement of expectations for everyday comforts, individuals these days have enough nourishment to bolster themselves and have higher need for diet. Individuals hope to know precise substance of their weight control plans and distinguish different sustenances that improve wellbeing and forestall potential maladies.

In this way, an extreme need which is forced on sustenance inquires about improvement. Nourishment comprises of an exceptionally intricate framework, and customary examination of instruments dependent on an investigation of constituents and parts are ineluctably being limited by the variables, for example, the extraction techniques.

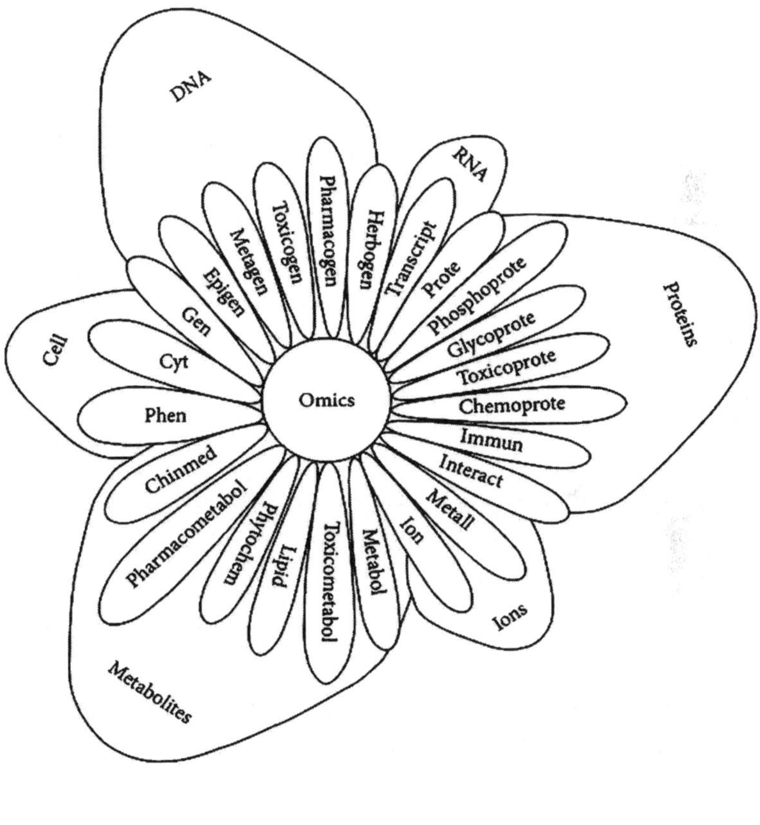

Figure 8.4: Omic-approaches and their area of study in systems biology.

Source: https://upload.wikimedia.org/wikipedia/commons/thumb/3/3d/Omic-approaches-and-their-area-of-study-in-systems-biology.jpg/580px-Omic-approaches-and-their-area-of-study-in-systems-biology.jpg

Moreover, from the standpoint of food nutrition, more pressure on specific ingredients often leads to neglect of other ingredients or components that impose or impact health. Capozzi and Bordoni in the year 2013 discussed the importance of taking into account the whole picture of all nutrients and their functioning on the entire human body.

Food-omics looks at high-throughput omics to examine the nature and safety of food safety and food nutrients, Capozzi and Bordoni, (2013). These omics systems have been facilitated by the complete sequencing and sequence studies of several genomes.

The study of food-omics first entered in the year 2007. In the year 2009, professor Alejandro Cifuentes from the Institute of Food Science Research which is the full form used in place of CIAL of the National Research Council of Spain first explained food-omics as a new discipline that studies the food and nutrition domains by using omics technologies, as reported in Journal of chromatography in the year 2009.

The aim of food-omics research is to improve consumers' well-being and health as stated by Capozzi and Bordoni in the year 2013, Herrero in the year 2010, Herrero in the year 2012 and Cifuentes in the year 2013. This field encompasses two domains: food science and nutrition science.

The following approaches allow for maintaining health and preventing diseases nutrigenomics, microbial genomics, toxicogenomic, nutritranscriptomics, nutriproteomics, nutrimetabolomics, and systems biology, food-omics to amalgamate agricultural products, food compositions, and dietary structure with individual physical trait differences , (Capozzi and Bordoni, 2013). The assessment of food ingredients, mechanisms of nutrition, food quality and processing have also been taken into account.

The concept of food-omics looks at food science development with an application of systems biology techniques on food nutrition and health studies.

Since the whole end analysis of human genome project which is the full form used in place of HGP, scientists have conducted researches on the physiological and biochemical mechanisms in vital processes of the aspects of genes including epigenome, single-nucleotide polymorphism SNP, etc., RNAs including mRNAs, miRNAs, and liRNAs, proteins, small molecule metabolites, and other aspects of the systems biology.

Human health includes aspects of genetics, the environment, and nutrients. These have become a center point in food science, and systems biology approaches as basic research tools. It is all-comprehensive to include physics, biology, chemistry, and medical science.

The concept of food-omics was well received by specialists and numerous scientists. Since 2009, International Conference on Food-omics has been held every two years in Italy, and the third conference was held in 2013.

In 2012, the Electrophoresis magazine published a special issue on "Food analysis in the postgenomic era: Food-omics as stated by Cifuentes 2012)," which elaborated the scope of researches and analysis approaches of food-omics. In May 2013, food omics was a topic of discussion at the 10th International Conference of Food Science and Technology (ICFST) at Jiangnan University, China while on November 2013, the sixth international symposium on recent advances in food analysis was held at Prague of Czech Republic with food-omics as one of the thematic forums. .

8.4 NUTRITIONAL SYSTEMS BIOLOGY MODELING: FROM MOLECULAR MECHANISMS TO PHYSIOLOGY

Nutritional science is currently undergoing a data boom as a constantly increasing number of researches are incorporating various methods from genomics, transcriptomics, proteomics, and metabolomics. Yet, the use of traditional nutritional techniques to analyze this high-dimensional data information is still unclear.

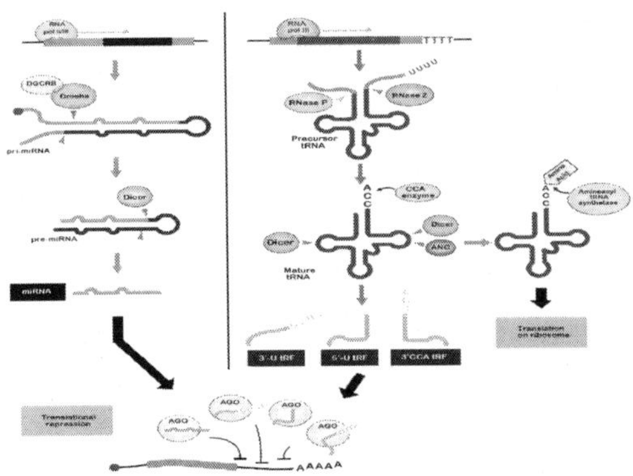

Figure 8.5: tRNA-Derived Short Non-coding RNA as Interacting Partners of Argonaute Proteins.

Source: https://live.staticflickr.com/757/21209969078_dedf519e41_b.jpg

A large section of computational modeling methods has been applied to wide-ranging stages of organization which are from molecules to humans. Several systemized systems include signaling and metabolic pathways, physiology and growth that each require specific computer tools.

This chapter discusses the role of computers and structures in food science and various successful representative modeling approaches.

8.4.1 Nutrition and Modeling

Nutrition research analyses "are the processes by which the living organism attains and utilizes the materials necessary for the maintenance of life and health", as defined by James S. McLester in his classic 1927 textbook. This involves the effects on an entire organism with each organism far away from thermodynamic equilibrium.

This requires the input of enough energy from surroundings in the form of nutrient combustion.. The overarching organizing principle expressed in the Dynamic Energy Budget theory, which considers that energy from food is extracted by the stomach, stored in reserves in the forms of body fat, protein, and carbohydrate pools, and distributed throughout the body to fuel the processes which are required for life.

These processes comprise the generation of heat, maintenance of components which is needed across cell membranes, the creation of gametes, the synthesis of structural mass, the development of maturity, somatic maintenance, and maturity maintenance. This allows for homeostasis by decoupling an organism's internal energy from the external world, important for nutritional physiology.

In contrast to the dietary macronutrient energy sources which are protein, carbohydrate, and fat, dietary micronutrients, notably mineral elements and vitamins, also play a definitive role for the overall health of the organism. Several deficiency diseases such as scurvy, beriberi, anemia, goiter, and cretinism are the manifestations of inadequate dietary micronutrients.

There is a third class of essential amino acids and the omega-3 and omega-6 fatty acids that are involved in energy synthesis, mass production and the formation of specific bioactive compounds.

Many health disorders are not caused by dietary lacks, but more generally from imbalances intake and utilization of nutrients. While there is consensus that complete nutrition can prevent various chronic diseases, understanding the health effects of specific nutritional compounds is extraordinarily complicated.

Initially, conveyance of a nutritional disturbance is very hard to power over significant time-frame periods with these regularly have generally unpretentious impacts over the time scales commonly researched when contrasted with pharmacological molecules whose impacts are distinguished on time scales from minutes to days or weeks.

Second, it is extremely hard to unwind the unmistakable bioactivity of a nutrient when it is provided in a foundation diet containing several other bioactive segments. Third, it tends to be hard to evaluate the bioavailability of the supplement, particularly at the degree of organs or cells. Commonly, compartmental modeling approaches are utilized to portray the ingestion, dispersion, and destruction of a supplement. Normal to the vast majority of these models is the abnormal state of total, where the body is sufficiently clarified by just a couple of areas. Together, these models have single goal to give a sound premise to the assurance of the wholesome need of the body, and for the requirements in prerequisites both locally for organs and at the entire body level for various micronutrients.

While such conventional demonstrating strategies have been exceptionally valuable, the genuine test for displaying in food is to help comprehend and soundly control the mind-boggling connection among nourishment and wellbeing, which is dictated by the coordinated multiscale answers to supplements, running from the entire body to subcellular phases of association and after some time sizes of minutes to years.

This difficulty is quite apparent from the problems that arise with current efforts to pinpoint the precise role of nutrition in the metabolic syndrome. At the long-timescale and whole-body level of organization, a prolonged period which ranges between weeks to years of consuming more energy than is expended results in the gradual development of obesity and increases one's risk for developing insulin resistance, a hallmark of the metabolic syndrome.

The research of insulin resistance has explained that the effect of this hormone on organs and tissues occurs on the time scale of minutes to hours. An example is the process of lipid and carbohydrate metabolism as stimulated by insulin.

Such molecular mechanisms have been associated with gene expression changes along with complex intracellular signal transduction and trafficking pathways and modification of enzyme kinetics ,. Furthermore, the manufacturing of insulin by pancreatic beta cells generates feedback to glucose and amino acids and can be manipulated by fatty acids, all of which can clearly be influenced by diet and nutrition. These cells display unique electrophysiological properties in terms of oscillation and bursting associated with insulin secretion on the time scales of seconds to minutes. Thus, understanding how nutrition affect the mechanisms underlying insulin resistance needs a quantitative analysis and description of a multiscale, being greatly coupled regulatory network that consists of thousands of components, having range over subcellular to whole-body levels of the organization and spanning time scales from within seconds to years.

Although a theoretical perspective as outlined above can be derived from literature without too much effort, it is quite difficult to develop an integrated quantitative understanding that wide ranges for the entire complexity of the mechanisms involved. In principle, mathematical structure and schematics offer this capability and therefore are needed to fully understand the physiological basis not only of the metabolic syndrome but of the role of nutrition in health and disease in general.

Without such an analysis and linkage method, it is non-ignorable that one will get lost in the web of bubbles and arrows quite typical of conceptual schematics and find oneself unable to weigh the relative importance of each component or interaction in finding out the overall physiological phenotype.

The section of mathematical modeling in nutrition is quite diverse and presently no single mathematical formalism makes one create the required linked quantitative understanding of nutrition as expressed and discussed above. Therefore, in developing researchers' vision of what is required in the coming years, the researchers now review numerous representative models that have successfully addressed key nutritional questions and together may help point the way to a more integrative modeling method.

First, researchers analyze modeling methods for processes at the cellular level explaining the biochemical processes that are signaling- and metabolic pathways that generate to convert food ingredients into energy and building structures for the cell as the fundamental unit of life. Insight into these processes teaches us how metabolism is being regulated at its most basic level. This is the entry point of the omics data.

Second, researchers conclude, the usage of metabolic flux analysis which is the full form used by MFA as a structure for the quantitative analysis of material fluxes within the single-cell as well as between different cell populations and organs, up to the whole-body level. Thus, metabolic flux analysis forms a natural connection amid different levels of organization and different time scales.

Thirdly, researchers' review compartmental models of lipoprotein metabolism, because lipoproteins are the major mediators of lipid trafficking amid organs, and many processes linked with lipids are related with diseases such as insulin resistance or cardiovascular diseases.

Finally, researchers stated mathematical models of body weight and composition regulation and the complex linked with macronutrient metabolism at the whole-body level. Modeling at this whole-body physiological level demonstrates the main need of considering long time scales that are characteristic of chronic diseases like obesity and metabolic syndrome.

Of course, researchers cannot cover all regions of modeling in the sector of nutrition in this section. For instance, researchers will not review schematics and procedures of gut-associated processes of nutrient absorption and bacterial conversion of nondigestible food constituents into such important compounds as short-chain fatty acids.

Another important region that researchers will not cover is the structure of the neurohormonal regulation of food intake and metabolism. Yet, the collection of models that researchers chose to analyze make sure that this chapter cover process occurring on a vast space-time spectrum, from nanometer to meter and from microseconds to years to come.

Therefore, the four regions of modeling that this chapter research provides a sufficiently wide basis for our aim, that is to analyze the available computational methods that are key to answering central questions in nutrition and that can serve as a platform for the development of more integrative systems models.

8.4.2 Cellular Modelling Approaches

Mathematical methods of cellular procedures can be used to simulate cellular behavior to better understand the complex structure underlying experimental observations. This understanding can highlight the reaction of a system when a gene is deleted or a nutrient is added.

Otherwise, the cellular structure may also deal with more general issues such as how control is distributed in a complex linkage. This can be followed by experimental analysis to examine the true behavior of the cell.

To allow such measurable simulation and prediction, cellular structuring ultimately aims at a detailed, mechanistic explanation of molecular processes developed in single cells. Towards this aim, known pathway structures are translated into differential equations, which after analysis of the unknown standards from experimental information can be used for dynamic simulations of a pathway or linkage behavior.

However, such mechanistic structure methods are presently only held maintainable for rather small course ways or links. Thus, even on the cellular level, there is a crack between kinetic mechanistic structures on the one hand and more coarse-grained modeling methods on the other hand. The latter is greater in scope that is they encompass more modeled components and interactions up to the genome-scale but explain the interactions between the modeled mechanisms with less mechanistic aspect.

8.4.3 Mechanistically Detailed Kinetic Models

The first kind of cellular models explains molecular mechanism at the cellular stage based on ordinary differential equations. These models comprise of balance equations explaining the dynamic concentration changes of the measured molecules with already suitable rate laws that is mass action or Michaelis-Menten kinetics.

In most of the cases, these structures consider only a few dozens of molecules and either aim on metabolic or on signaling procedure. Prominent instances of such schematics include a model of glucose metabolism of the red blood cell and a model for the yeast *Saccharomyces cerevisiae* that automatically explains the organism›s response to osmotic shock.

The latter deals with the thermodynamics of stimulation of a receptor followed by mitogen-activated protein kinase cascade dynamics, activation of gene expression, and the adaptation of cellular metabolism. As such, this system is one of the few kinetic models that explain procedures on more than one cellular level.

A present model of the carbon transfers in the hepatic folate cycle is an instance of a model directly in relation to a specific nutrient. More cellular models can be found in dedicated model repositories. From such kinetic models, mechanistic vision about the modeled molecular linkages can be originated by means of numerical manipulation and other computational analyses such as metabolic control analysis, which figures out how the control of flux is distributed in metabolic networks.

One of the major challenges in the growth of these mechanistic models is that varied uncertainties exist in the molecular mechanisms and in the respective model parameter levels. This usually involves *in vitro* measurements systems or dynamic measurements of *in vivo* systems.

Although currently such data are being generated for single-cell organisms, model structure and parameter identification challenges remain huge, particularly with models of larger size.

8.5 THE CHALLENGES FOR MOLECULAR NUTRITION RESEARCH

Nutritional systems biology may be defined as the goal of molecular nutrition research, where all relevant aspects of the regulation of metabolism in health and disease states at all levels of its complexity are taken into account to describe the molecular physiology of nutritional processes. The complexity spans from intracellular to inter-organ dynamics, and involves iterations between mathematical modeling and analysis employing all profiling methods and other biological read-outs.

Based on such dynamic models, we should better understand how the nutritional status and nutritional challenges affect human metabolism and health. While this looks impossible, these topics are being addressed.

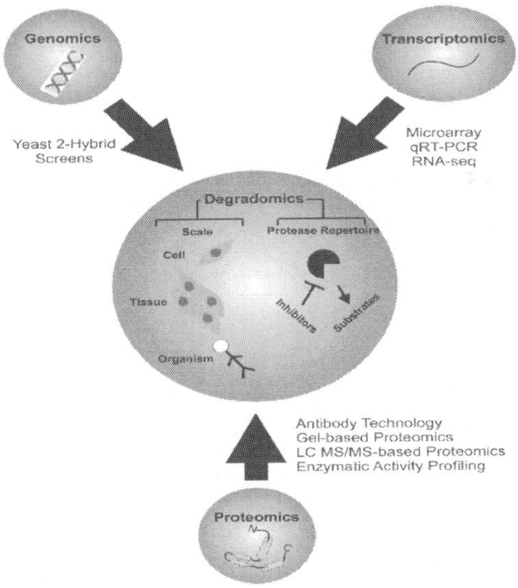

Figure 8.6: Degradomic Figure.

Source: https://upload.wikimedia.org/wikipedia/commons/a/ac/Degradomic_Figure_Wikipedia.png

This section delivers examples and some recommendations for the molecular nutrition research region to move onto the different systems level.

Let's start with a quote: "Perhaps surprisingly, a concise definition of systems biology that most of us can agree upon has yet to emerge" , Ruedi Aebersold, Ph.D., a faculty member of the Institute for Systems Biology, Seattle, USA.

Since there is no uniting theory of what systems biology is, researchers may take the liberty here to explain what researchers consider as Nutritional Systems Biology which is the full form used in place of NSB.

"Nutritional Systems Biology shelters all different methods being aimed towards understanding the chief processes that describe nutrition in the statement of regulation of transcription, protein synthesis and turn-over, metabolism and genome constancy in health and disease states at all levels of its complexity, to simulate these processes and to predict the outcome of any alteration whether genetic or nutritional of the system.

NSB starts at the molecular and cellular levels and takes the regulatory processes onto the level of the different organs and the inter-organ flow of information, nutrients and metabolites to describe most comprehensively the organism response and the processes quantitatively in mathematical terms".

8.6 PROTEOMICS AND THE RELATION BETWEEN NUTRITIONAL RESEARCH

If Deoxyribonucleic Acid (DNA) is the blueprint for life, then proteins are the cell structure. The genes in Deoxyribonucleic Acid are being transformed into proteins, strings of amino acids that fold into three-dimensional structures. The kind and order of the amino acids in a protein will transform its shape and determine its special function. Proteins are the molecules that make life happen, they are the power plants that turn food into energy, the machines that make cells transmission, and even the computers that read Deoxyribonucleic Acid to make more proteins.

The data to build every protein in an organism is contained in the DNA, but not every protein is being generated at once or in the same quantity. Think of a cell in human liver and a cell in your retina – both cells contain identical DNA, but very dissimilar subsets of proteins are being generated in order to give each cell its special function.

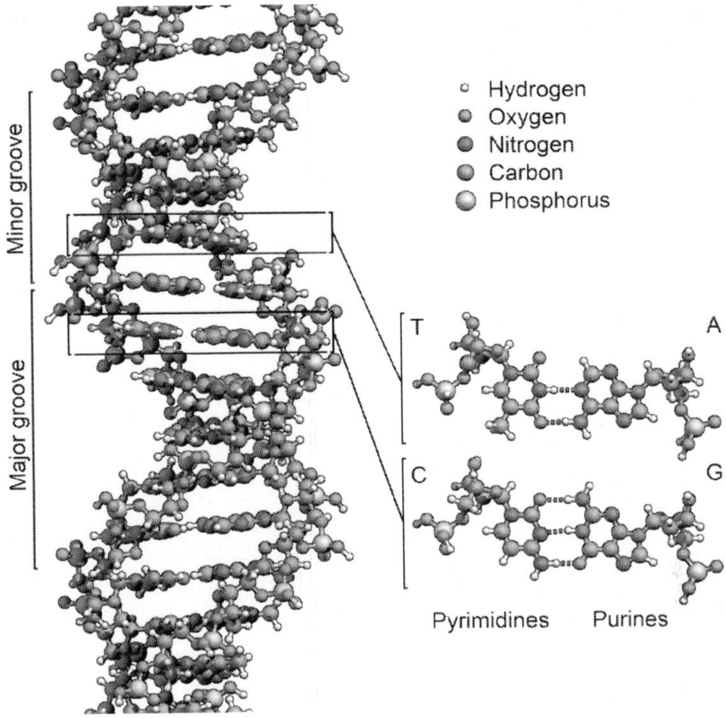

Figure 8.7: Deoxyribonucleic Acid structure.

Source: https://upload.wikimedia.org/wikipedia/commons/4/4c/DNA_Structure%2BKey%2BLabelled.pn_NoBB.png

Proteomics is the discipline of detecting and analyzing the proteins being available in an organism. At ISB, researchers use state-of-the-art technical instruments and cutting-edge computational methods to detect thousands of proteins at once, giving us a systems-level projection of the molecular structure of life.

8.7 PROTEOMICS AS A TOOL FOR THE MODELING OF BIOLOGICAL PROCESSES AND BIOMARKER DEVELOPMENT IN NUTRITION RESEARCH

Nutrition research has constantly been started to adopt the proteomics approaches to analyze changes in the protein complement of a biological system. This makes modeling of biological processes in answer to dietary interventions, as well as the elucidation of novel biomarkers for health or disease that are quite normal to such involvements.

There are limited researches on the effect of micronutrients on the proteome, so this section has its main attention to rather more on dietary intervention researchers that have used proteomics mainly classical two-dimensional gel electrophoresis combined with mass spectrometry to elucidate variation in course ways that relate to glucose and fatty acid metabolism, oxidative stress, antioxidant defense mechanisms and redox status.

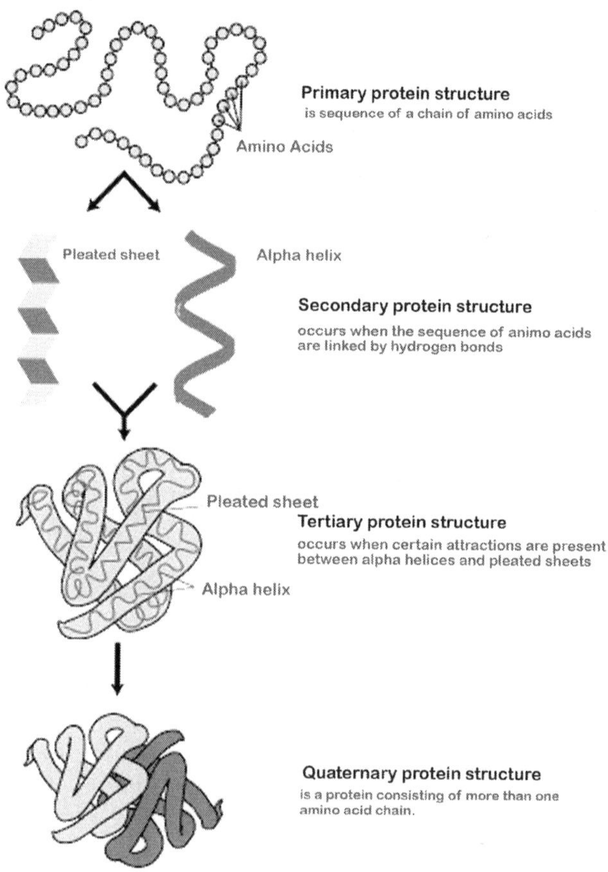

Figure 8.8: Protein structure.

Source: https://upload.wikimedia.org/wikipedia/commons/a/a6/Protein-structure.png

The main ability to analyze regulation of low plentiful proteins, such as those being involved in inflammatory course ways, as well as the assessment and validation of newly being discovered candidate biomarkers in human biofluids, may usually depend on the introduction of more analytical and sensitive approaches like multiple reaction monitoring (MRM) and multiplexed immunoassays in nutrition research.

Epidemiological studies point out at the role of diet in disease and its incidence and severity. Yet the detection and mechanisms of the bioactive components in the diet or food is challenging. Diets are considered complex with several food items together as a complex mix.

Their connections with physiological systems, therefore, are quite impossible to deduce using traditional experimental approaches. This requires the global nature of nutrigenomics to address all these aspects.

Also, knowledge about the actions of foods and nutrients on all possible physiological outcomes in relation to, for instance, chronic disease development, is quite needed to identify dietary strategies that will endorse health and delay the onset of chronic diseases.

The potential value of proteomics for nutritional sciences has been documented for some years and many researchers have written excellent researches, for instance promulgating its use in nutrition. Though, the total number of researches published up to the year 2007 surpasses the number of research kinds of literature on proteomics!

Thus, whilst proteomics is being projected as an emerging and promising field that allows the elucidation of systems of action, as well as growth of relevant biomarkers of health or disease, the main usage of this technique in dietary intervention trials, is still rather quite limited. In relation to micronutrient metabolism, there are only a few papers that investigative either the proteomic as a source of identifying dietary treatment or which highlights the possible use of proteomics to be used to identify biomarkers of health status.

In this section, therefore, researches have also included other dietary behaviors that can be used as examples of the approaches and hopefully deliver inspiration to others to adopt the techniques in their micronutrient research.

8.7.1 Proteomics as a Tool to Elucidate Mechanisms of Action

Proteomics has the benefit over complementary Deoxyribonucleic Acid (cDNA) micro-arrays in that it evaluates the functional product (which is protein) of gene expression to permit analyzing the changes connected to the activation or inactivation of proteins with the help of dietary interventions.

These types of proteins can not only play important physiological role in the target organs but could also reflect alterations in components which is originated with the help of the dietary intervention when spread in the blood.

Classical two-dimensional (2D) gel electrophoresis, when combined with the spot test analysis with the help of mass spectrometry, is most extensively utilized with respect to other technical methods in the field of proteomics to distinguish the changes in the individual proteins of tissues, cells as well as biofluids upon the intervention of nutrition.

On the other hand, this approach is one of the most labor-intensive of the various form of two-dimensional 2D separation approaches which are available, it actually produces a physical separation of complete polypeptides from each other, giving the information about the weight of the molecule and iso-electric point, limitations that can be utilized to narrow down the evaluation of the protein.

In the meantime, the dynamic range of protein concentrations spans six orders of magnitude in a cell, and even ten 10 in the body of a human (for instance in plasma). Proteins that take part in the inflammatory pathways, such as cytokines, are secreted to move around the body at a concentration nine orders of magnitude less than albumin. It is at present very complicated to imagine and identify various types of regulation of low abundant, lacking affinity for water, acidic or basic proteins.

Consequently, two-dimensional 2D gel electrophoresis may not always represent the most effective set of instruments to disclose the influences of nutritional interventions on inflammatory pathways. On the other hand, two-dimensional 2D gel electrophoresis and mass spectrometry have developed itself as a well-appreciated set of instruments in explaining the alterations in the pathways that narrate to glucose and fatty acid metabolism as well as the pathways which are connecting to oxidative stress, antioxidant defense devices and redox status.

Studies in the laboratory have explained that two physically alike linoleic acid (CLA) isomers had different operational impacts on the atherosclerosis development and insulin resistance in Apoe − / − mice.

Utilization of trans10, cis12 CLA, a fatty acid present in additions that are at present easily obtainable, was pro-atherogenic and encouraged trails which took participation in the growth of insulin resistance, as demonstrated with the help of the instantaneous alterations in several types of enzymes which took participation in the gluconeogenesis, glycolysis and ketogenesis.

Utilization of the cis9, trans11 isomer of CLA, on the other hand, was anti-atherogenic and more particularly encouraged heat shock protein (HSP) 70 kD protein levels. Even though one may reflect this to be a sign of amplified oxidative pressure, it has been postulated that up-regulation of HSP 70 kD protein upon exposure to a comparatively mild stress stimulus (like cis9, trans11 CLA) could defend in contradiction of exposure to more simple stress later, maybe together with the anti-inflammatory pathways (mediated with the help of NF-κB). Certainly, amplified levels of human serum HSP 70 have been linked with a low coronary artery disease risk, and it is not dependent on traditional risk factors.

REFERENCES

1. Auley, M. (2013). Nutrition Research and the Impact of Computational Systems Biology. *Journal of Computer Science & Systems Biology*, 06(05).
2. Badimon, L., Vilahur, G. and Padro, T. (2016). Systems biology approaches to understand the effects of nutrition and promote health. *British Journal of Clinical Pharmacology*, [online] 83(1), pp.38-45. Available at: https://bpspubs.onlinelibrary.wiley.com/doi/pdf/10.1111/bcp.12965 [Accessed 8 Jun. 2019].
3. de Graaf, A., Freidig, A., De Roos, B., Jamshidi, N., Heinemann, M., Rullmann, J., Hall, K., Adiels, M. and van Ommen, B. (2009). Nutritional Systems Biology Modeling: From Molecular Mechanisms to Physiology. *PLoS Computational Biology*, [online] 5(11), p.e1000554. Available at: https://journals.plos.org/ploscompbiol/article?id=10.1371/journal.pcbi.1000554 [Accessed 8 Jun. 2019].
4. metabolomicsconference. (2019). *Best Metabolomics Conferences 2019 | Biochemistry Conferences USA | European Metabolomics Events | Middle East | Russia | Asia | Japan*. [online] Europe. metabolomicsconference.com. Available at: https://europe.metabolomicsconference.com/events-list/transcriptomics-and-proteomics [Accessed 8 Jun. 2019].
5. Moore, J. and Weeks, M. (2011). Proteomics and Systems Biology: Current and Future Applications in the Nutritional Sciences. *Advances in Nutrition*, [online] 2(4), pp.355-364. Available at: https://www.ncbi.nlm.nih.gov/pmc/articles/PMC3125684/ [Accessed 8 Jun. 2019].
6. Moore, J. and Weeks, M. (2011). Proteomics and Systems Biology: Current and Future Applications in the Nutritional Sciences. *Advances in Nutrition*, [online] 2(4), pp.355-364. Available at: https://www.researchgate.net/publication/221830284_Proteomics_and_Systems_Biology_Current_and_Future_Applications_in_the_Nutritional_Sciences [Accessed 8 Jun. 2019].
7. Norheim, F., Gjelstad, I., Hjorth, M., Vinknes, K., Langleite, T., Holen, T., Jensen, J., Dalen, K., Karlsen, A., Kielland, A., Rustan, A. and Drevon, C. (2012). Molecular Nutrition Research—The Modern Way of Performing Nutritional Science. *Nutrients*, [online] 4(12), pp.1898-1944. Available at: http://www.scitcentral.com/journals.php?journal=53 [Accessed 8 Jun. 2019].

8. Nutrition | Case Western Reserve University. (2019). *PhD Program in Systems Biology and Bioinformatics (SYBB) | Nutrition | Case Western Reserve University*. [online] Available at: https://case.edu/medicine/nutrition/education-programs/phd-programs-nutrition-and-systems-biology-bioinformatics/sybb-phd-program [Accessed 8 Jun. 2019].
9. Panagiotou, G. and Nielsen, J. (2009). Nutritional Systems Biology: Definitions and Approaches. *Annual Review of Nutrition*, 29(1), pp.329-339.
10. Panagiotou, G. and Nielsen, J. (2009). Nutritional Systems Biology: Definitions and Approaches. *Annual Review of Nutrition*, [online] 29(1), pp.329-339. Available at: https://www.annualreviews.org/doi/abs/10.1146/annurev-nutr-080508-141138 [Accessed 8 Jun. 2019].
11. Moore, J. and Weeks, M. (2011). Proteomics and Systems Biology: Current and Future Applications in the Nutritional Sciences. *Advances in Nutrition*, [online] 2(4), pp.355-364. Available at: https://www.researchgate.net/publication/221830284_Proteomics_and_Systems_Biology_Current_and_Future_Applications_in_the_Nutritional_Sciences [Accessed 8 Jun. 2019].
12. Norheim, F., Gjelstad, I., Hjorth, M., Vinknes, K., Langleite, T., Holen, T., Jensen, J., Dalen, K., Karlsen, A., Kielland, A., Rustan, A. and Drevon, C. (2012). Molecular Nutrition Research—The Modern Way of Performing Nutritional Science. *Nutrients*, [online] 4(12), pp.1898-1944. Available at: http://www.scitcentral.com/journals.php?journal=53 [Accessed 8 Jun. 2019].
13. Nutrition | Case Western Reserve University. (2019). *PhD Program in Systems Biology and Bioinformatics (SYBB) | Nutrition | Case Western Reserve University*. [online] Available at: https://case.edu/medicine/nutrition/education-programs/phd-programs-nutrition-and-systems-biology-bioinformatics/sybb-phd-program [Accessed 8 Jun. 2019].
14. Panagiotou, G. and Nielsen, J. (2009). Nutritional Systems Biology: Definitions and Approaches. *Annual Review of Nutrition*, 29(1), pp.329-339.
15. Panagiotou, G. and Nielsen, J. (2009). Nutritional Systems Biology: Definitions and Approaches. *Annual Review of Nutrition*, [online] 29(1), pp.329-339. Available at: https://www.annualreviews.org/doi/abs/10.1146/annurev-nutr-080508-141138 [Accessed 8 Jun. 2019].

16. systemsbiology.org (2019). *Proteomics · Institute for Systems Biology*. [online] Institute for Systems Biology. Available at: https://systemsbiology.org/research/proteomics/ [Accessed 8 Jun. 2019].
17. van Ommen, B., Cavallieri, D., Roche, H., Klein, U. and Daniel, H. (2008). The challenges for molecular nutrition research 4: the "nutritional systems biology level". *Genes & Nutrition*, [online] 3(3-4), pp.107-113. Available at: https://link.springer.com/article/10.1007/s12263-008-0090-5 [Accessed 8 Jun. 2019].
18. www.proteomicsconference.com. (2019). *Proteomics and Bioinformatics*. [online] Available at: https://www.proteomicsconference.com/events-list/proteomics-in-computational-systems-biology [Accessed 8 Jun. 2019].
19. Zhao, Y., Barrere-Cain, R. and Yang, X. (2015). Nutritional systems biology of type 2 diabetes. *Genes & Nutrition*, [online] 10(5), p.2. Available at: https://genesandnutrition.biomedcentral.com/articles/10.1007/s12263-015-0481-3 [Accessed 8 Jun. 2019].
20. Zheng, C. and Chen, A. (2014). System Biological Research on Food Quality for Personalized Nutrition and Health Using Foodomics Techniques: A Review. *Journal of Food and Nutrition Research*, [online] 2(9), pp.608-616. Available at: http://pubs.sciepub.com/jfnr/2/9/13/index.html [Accessed 8 Jun. 2019].
21. Wu, Y., Mirzaei, M., Pascovici, D., Haynes, P. and Atwell, B. (2019). Proteomes of Leaf-Growing Zones in Rice Genotypes with Contrasting Drought Tolerance. *PROTEOMICS*, [online] 19(9), p.1800310. Available at: https://onlinelibrary.wiley.com/doi/abs/10.1002/pmic.201800310 [Accessed 8 Jun. 2019].

9

Application and Uses of Proteomics in Food Science and Technology

CONTENTS

The Proteomics field is very wide, and its application is in variety of fields like food science, in area of medicine and other fields. It is mainly the study of breaking down of protein into smaller units or new type of proteins. There are many types of bacteria that can cause food poisoning and proteomics can be applied to fight those bacteria. It is used as a tool to develop and preserve food in such a way that that food life is increased.

Proteomics has very wide application in preservation of meat. It is applied in meat science to preserve the meat from getting decomposed as meat is an essential part of diet of billions of people around the globe.

In future the application of proteomics in field of food processing is bound to grow.

9.1 INTRODUCTION

The classification of proteins in cow milk and in the latest dietary foods like breast milk alternatives is due to the implementation of proteomic technology. The food allergy that is very much widespread in children is milk protein found in cow milk, that affects around three percent of children in the starting three years of life. The appropriate cow's milk antigens that are found in very less amount in human milk is caseins, and foreign substances that are in very less amount is lactoglobulin.

There is a growing interest in getting milk from animals that does not come from cattle category like goats, donkeys and camels. The human babies that are intolerant to mothers' milk can drink donkey milk as demonstrated in clinical tests but is not yet clear what exactly is the present in donkey milk. It has been discovered by integrating diverse electrophoretic methods and MALDI-TOF MS that donkey milk is generally devoid of s2-casein and k-casein, it is the reason for its natural hypo-allergenicity.

Studying Bioactive Plant Components

The large variety of secondary metabolites or phytochemicals of less molecular weight in less amounts are found in plant food. The protection from pathogens is provided by these compounds, and from growth controllers and pigments. Vegetarian food have become famous in western countries due to its health benefits whereas Indian food is generally full of vegetables and fruits as there is a large majority of people who consume vegetarian food.

Vegetarian food has become famous due to its health benefits only over the last ten years. The consumption of fruits and vegetables make immunity stronger against serious diseases, such as cardiovascular disease and cancer that are associated to the existence of phytochemicals that has been shown by a large amount of evidence in favor of sanitation practices. The restrictions to get a full proteome extract as the result of the less protein found inside it, the high-water level and the existence of foreign materials are the challenge put up by the breakdown of proteins in fruit and vegetable.

The red color of fruits of plants that are related to the category of the citrus fruits is due to the accumulation of large amounts of anthocyanins that differentiates it from others. The integration of 2DE and LC-MSMS methods and abstraction of protein with phenol differentially expressed proteins can be recognized and assembled into different efficient sets.

A large amount of differently stated proteins in Moro might be associated, though not directly connected to, to the synthesis of anthocyanins can be found by guessing on the biological importance of such differences. The growth in appearance of sugar metabolism enzymes, along with transport proteins, can fulfill the need of carbon structures for anthocyanins synthesis. The attack by pathogens is responsible for combination of defense proteins in large amount and that can be associated with anthocyanins. On the other hand, comparing proteomic data with previously obtained transcriptomic results, discrepancies were found to arise from the fact that proteins are subjected to post-translational processing and degradation controls, modulating their temporal expression, location, and efficiency in tissues.

Meat Quality

The biological and environmental factors communication decides the meat quality, and between genome and quality, the molecular connection is proteome. The procedure that is used for tenderization is very important for betterment of meat quality in the property of meat quality. The building of tenderness is reliant on the shape and integrity of skeletal muscle cell and on procedures that alter those proteins and their connections, mainly credited to endogenous proteolytic enzyme systems, largely too reliant on calcium non-lysosomal micro-calpains.

The meat tenderization is done by proteins that are targets in this process. There has been an observation that the dissolving capacity of some cytoplasmic protein HSP and glycolytic enzymes decrease, few full-length proteins are produced from cellular structures like membranes and myofibrillar network produces few full-length proteins and some proteins are broken down that is glycolytic enzymes, structural proteins at the time of maturity of muscles.

For example, the presence of titin (a large myofibril cytoskeleton protein disintegration product) moving at 1200 kDa has been accepted in myofibrils from preserve cow meat that had lower cut off force and more needed softness scores. The softness of pig meat pork has been observed to have connected to the deposition of the main major myofibrillar protein pieces, actin from 44 to 28 kDa and myosin with speed 49 kDa.

There has been association between softness and the searching of pieces of bovine myosin heavy chain. Even though only a minor part of actin and myosin heavy chain is degraded in the postmortem, the very less breaking down of protein causes weakening of structures and then tenderization yet only the small part of actin and myosin heavy chain is destroyed.

The breaking down of the myofibrillar protein troponin T and the resultant presence of a 30kDa piece has been found to be a good sign of the level of complete post-mortem break down of protein and an indicator of meat softness. The experts have stated that there has been the association between softness and the plenty of proteins of mitochondrial membrane that is prohibiting, mitofilin, elongation factor Tu, voltage-dependent anion-selective channel protein 1 and protein 2, NADH-ubiquinone oxidoreductase.

The implementation of process of apoptosis that is catalyzed d by proteases caspasesis frequently present. . In addition, 31 kDa fragment of actin found overrepresented in tender meat may also support the apoptosis hypothesis because actin is known as a target of caspases and the fragment is used as an apoptosis marker.

The disintegration of cytosolic enzymes produces peptides that can be useful indicators of proteolytic activity and consequently the good indicator of the quality of meat. The degree of triosephosphate isomerase has been observed not related in any way to meat cutting down force, signifying that post-mortem glycolysis can act in an important manner in post-mortem softening.

The conventional products that are always in popular demand in western countries are sun-dried ham and fermented sausage, made from pig and beef fat and meat. The influencing of final features of in context of taste and consistency is caused by long time of ripening and severe breaking down of protein. In dry-cured ham, the severe breaking down of protein is persisted by endogenous exo- and endopeptidases that are mainly of lysosomal origin, namely cathepsins that work for a long time and produces free amino acids and peptides.

In samples of meat, the task of proteolysis varies broadly among meat samples and is clearly affected by standards of processing that is salt, temperature and time, giving rise to different patterns of proteolytic parts. The categorization of sun-dried ham is made possible by biochemical markers and is recognized by proper information given by protein fragmentation.

There are few researchers and experts that have conducted studies on proteomic effects on meat products, there are experts that have said the bitterness in Italian aged hams is due to a high content of lipophilic amino acids and oligopeptides. In fresh pig meat, there is a high activity of cathepsin B that is associated with the creation of protein chains in aged ham.

The possible biochemical assessment of bitter taste development can be determined by measurement of cathepsin B. This type of association is also has been observed in other pig meats, that is also produced by same methods of longer ripening. The breakdown of myofibrillar protein into fourteen naturally originating small peptides has been recognized to contribute to the generation of taste in this variety of pig meat.

A proteomic study of 2DE followed by MALDI- TOF reported an evolution of protein map fragmentation throughout the ripening process of Italian dry-cured hams, showing a massive disappearance of both sarcoplasmic and myofibrillar proteins. Along with intense proteolysis, a high amount of a 135 kDa proteolytic fragment recognized by immunodetection and MALDI-TOF as derived from myosin heavy chain 1 was found.

Such a degradation pattern strongly resembles that found by *in vitro* incubation of native myosin with muscle specific cathepsin B, which yielded a similar fragment of C-term coiled-coil tail of myosin heavy chain.

9.2 PROTEOMICS AS A TOOL FOR PRODUCT AND PROCESS VALIDATION AND OPTIMIZATION

The organized proteomic examination along a plant-scale wet fractionation procedure of alfalfa biomass is carried out by experts. The chemical variations, heat-shock protein responses and proteolytic destruction is started by the production process. It was also demonstrated that during biomass processing, especially thermal treatment, a certain amount of cellular instructions is kept safe such as the initiation of heat shock and reduce stress proteins, at the time of processing of biomass as stated by experts.

The packing and storage and the processing process of protein-rich food of animal origin such as porcine meat start other modifications in meat and proteolytic degradation of structural proteins.

The development in process of food processing has caused the decrease in the time taken during production and brought an improvement in product quality. The requirement for better process control is caused by an increase in capacity of manufacturing. The modification in standards during processing and production process in the milk or meat processing industry is now very easy due to the use of software-led computer control systems. The features of many food products through the process of manufacturing is attributed to the type of proteins present in the final product.

The quantity of protein present in the product and its collection is accountable for physicochemical properties, like the thickness, resistance to heat and vapor pressure, nourishing and sensual properties of milk, meat and products made from grains. The gluten proteins are responsible for ideal features products made from wheat flour, and casein is the major protein in milk and milk-based products. The proteins and protein-based groups are properly categorized due to their importance.

The alteration in physicochemical properties at the time of production depends on more than one protein found in suitable quantity as protein structures of other foods like meat and meat products, and fruit and vegetables are tougher.

The consistency of meat and juiciness are the most essential of all taste features fulfilling their quality a large amount of pig and buffalo meat are eaten fresh by the public. The myosin, actin, desmin, and tubulin as structural proteins are responsible for the softness of meat as stated by research into proteomics.

The sudden modification is the introduction of some proteins during the starting phases of the process. Inappropriate heat treatment of milk, meat, cereal products or fruits and vegetables can negatively influence the product quality. The protein denaturation and the intrinsic range of chemical reactions are the major alterations that are also known as Maillard reaction.

In the latest study, the further revelations about Maillard reaction, in the context of proteomic application has been made. There is large number of features of food products such as color, consistency, the speed and easily it can digest, and nutritional value are impacted by the Maillard reaction.

The food nutritional value might be affected as the result of linking the essential amino acid such as the side amino group of lysine. The quality and nutrition value can be affected by any more modifications in meat and meat products via by glycation of proteins. In the Maillard reaction it is the first step. The contents of food, processing and storage situations can be altered to control the reaction.

The presence of dangerous acrylamide in food at the time of roasting, toasting and process of frying the food is caused by Maillard reaction between amino acids, primarily asparagine residues and falling sugars such as fructose, galactose, lactose and glucose.

The trigger of allergy from milk products can be caused by carbonylation of milk proteins such as b-lactoglobulin at the time of industrial treatment. The immunoblot can find out carbonylated proteins and recognized by MALDI-TOF MS or electrospray ionization tandem mass spectrometry after electrophoretic separation and in-gel digestion. The carbonyl attached with protein content in heat-treated milk samples was positively associated with the strictness of the treatment in the baked product to get the smell and to better the physicochemical features of whey proteins the Maillard reaction can be started to gain these objectives.

There is another form of chemical breakdown of proteins known as deamidation. The glutamine or asparagine residues are hydrolyzed to convert into glutamic acid or aspartic acid, which is an irreversible reaction. The identification of this form of protein conversion is possible by using mass spectrometric methods. More changes can be expected during the manufacturing process due to posttranslational alterations of proteins. The heat-sensitive phosphorylated serine and threonine residues can produce dehydroalanine and methyl-dehydroalanine, correspondingly.

The more synthetic products, such as lysinoalanine, lanthionine, and histidinoalanine can be produced by cross-reaction of many types of amino acids. The difference in ability is to dissolve of food proteins, that is wheat glutenins, mainly show its capacity to form inter- or intramolecular disulfide bonds.

The latest innovated online liquid chromatography-mass spectrometry with electron-transfer dissociation is a trustworthy process to find out disulfide connections before and at the time of processing of protein mixtures.

9.3 FOOD-BORNE PATHOGEN AND ALLERGY

The unpleasant response by the human body to foreign or sometimes internal metabolism, reasons or elements is normally defined as allergy. The response of the to the body is triggered by allergen. The allergy in the human body is normally due to growing immunoglobulin, increased histamines, cytokines which are produced in response to antigen or allergen. The immunoglobulin is the reason behind food allergic diseases.

There been an increase in the need of molecular methods to diagnose the allergen and body response associated with food-based allergy. There are detailed research works that have been conducted to identify the allergen in plants and animals' proteins. The researchers have documented the main food born allergens like in fishes par albumin, in cattle milk casein, Tropomyosins from Crustaceans and mollusks and Bet v 1 associated proteins in both monocot and dicot plants.

The growth of responsive detection and quantification methods are essential for allergen diagnosis, therapy, and risk assessment and for establishing current legislation are the reason behind the detailed documentation of proteomic research in the field of food allergens. The recognition of allergic diseases or allergenic proteins and genes is very essential. The international planning of health organizations is increasingly based on the idea of food quality and safety and is an impact on health. The spoiling of food is generally caused by microbes or enzymes.

The health of people gets affected by the spoilage of food and it also impacts the social, economic and public health importance. The recognition of molecular markers for food spoilage or pathogenic micro-organisms is essential for the execution of proteomics for examining the nutritional quality of food. The diverse types of microorganisms and complex substrates are found in fermented food, the proteome or metabolome of the starter culture can guess the quality of fermented food end-product. The change in the normal functioning of the body triggers disease, unusual conditions or even death by the microorganisms such as bacteria or fungi, sometimes viruses, prions, parasites and protozoa.

Microorganisms may affect the food in later stages but primarily they come in food during before the harvest, processing or after harvesting periods. The wrong method of storage or food storage before eating increases the risk of infection. The food poisoning, food infection and infection by poison are three types of food infection.

The bacterial or fungal poisons are released build or inside the body and are opposed to too many processing methods followed in food preparation. In 2016, Martinović has done a study on proteomics to fight such infections. He paid attention to factors in food pathogen and toxin identification by utilizing different proteomic methods.

The proteomics and other methods should be implemented to find out the shiga poison produced by *E. coli* O104: H4 and is behind large-scale food-based poisoning that was eliminated in Germany and France. In 2013, Aberg proposed the use of large scale spectrometric based methods for identification of toxins in different sources containing foodborne protein-based poison.

In 2015, Piras recognized the immunoreactive proteins of *Mycobacterium avium* subcategory. Paratuberculosis using proteomic based study. The spectra bank containing peptide large scale fingerprinting of various bacterial species that can support in future research related to proteomics for finding out pathogen (Böhme ,2012).

The pathogenesis is defined as an environment in which the microorganisms meet with host cell and cell organelles and cause changes in them. The proteins are produced during this procedure, that is irregular or up and down-regulated that is recognized on proteomic assessment and is getting popular in early diagnosis of diseases in food, animals and combined to find implementation in the field of food technology and biotechnology. In 2013, Giacometti stated few of the important protein secretions of mycotoxin and sea-based poisons. The different metabolic roads of microbial pathogenesis and plans for recognition of biomarker and resistances developed to biotoxin has been explained.

The implementation of proteomics in learning about the contact between the host and pathogen will be supportive in the innovation of new aim to make sure of food safety. The progress in area of animal production, agriculture, food processing and storage and for betterment in human health the progress in latest methods for recognition, monitoring and examining of foodborne dangers during production, processing and storage.

Adulteration is defined as the addition or subtraction of precious component in food with lower grade foreign product. For example, the adulteration of milk with water and starch, manufacturing of synthetic milk. The practice of selling low-value product in place of high-value product is called substitution. For example, selling of beef or carabeef as mutton.

The food varieties like food grains, pulses, milk, meat and other foods are either adulterated or its duplicates are being sold in the market and it has seen an increase in these types of crimes. The latest tools and methods are needed to solve this problem. To identify the adulteration people generally employ molecular methods but latest developments in proteomics have come out as the option.

In 2016, Garrido documented and recognized by utilizing shotgun proteomics analyses by medium and small enterprises, multiplexed, low and high-collision energy, data-independent acquisition methods the adulteration of whey protein with plant-based protein sources such as soybean, wheat and rice.

In 2002, Koller has done the categorization of rice by proteomics, *Oryza sativa* leaf, root, and seed tissue recognized over 2500 proteins, that is the most detailed proteome findings. Less expense in manufacturing has increased the rate of adulteration.

The whey protein obtained from cattle animals has high bioavailability and increased cost, while the products based on plants are normal side products obtained in the course of the harvest of crops. The health of humans is in danger and people also get cheated for hard-earned money from their pockets.

9.4 PROTEOMICS IN FOOD AUTHENTICITY

Food safety and quality are one of the main areas tangled in the certification of food. The government has implemented many controls in place to assure correct information and to evade exchange of species. The assessment of morphology or anatomy is used to classify a variety of food. The varieties that are closely associated are difficult to categorize, mainly the processed food products.

The quality of product and protection of customer can be ensured by utilizing molecular identification process that gives officials and food industries with the methods needed to conform with the need for labeling and traceability. There is a requirement for identification and many technologies have been developed for this task like electrophoretic, immunological and DNA methods in the past ten years.

The stability is not found in some proteins at the time of food processing and the labor and the hours needed for these processes are the disadvantages of the classical protein-based process. Also, the DNA-based process is not excused from some important boundaries.

There is a production of hydrolytic enzymes by a disturbance in cellular integrity at the time of processing of food products. The structure of DNA is affected in a wrong way, decreasing the length of fragments to be improved and as a result increasing the possibility of having not any certification are caused by integration of these enzymes with heat treatment and an acidic environment.

The methods of proteomics have been presented as futuristic technology for the purpose of food certification. Proteomics tools take advantage of the high-throughput capacity of MS to achieve fast, robust and sensitive protein and the categorization of peptide, recognition, and quantification can be achieved by taking benefit of the high-output capacity of MS by proteomics methods.

The varieties that are not properly categorized in genes database, evading the DNA amplification and sequencing that takes time on which methods can be implemented. Additionally, procedures that are based on proteomics can be computerized to manufacture fast, reproducible results that permit the increased output examination of products.

9.4.1 Discovery Phase

The discovery of a precise proteome broadly is the aim of the discovery phase, by utilization of reference pieces, to recognize possible animals' peptide biomarkers. The identification and the characterization of these diagnostic peptides are commonly performed using the bottom-up proteomics are utilized to recognize and classify the identified protein.

The proteins that are found varieties like peptides from parvalbumin and arginine kinase proteins have been shown to be good markers for the recognition of fish and shellfish varieties. The peptide biomarkers are classified, and it is the first step in the direction of resultant use as complex, diagnostic targets in the next phase of the pipeline.

9.4.2 Target-Driven Phase

The targeted proteomics methods are used to observe the species-specific peptide biomarkers that are categorized, for variety without sequenced genomes, the picking of peptides that are particular to species needs an earlier in-depth MS sequencing analysis in the second phase. The recognition of peptide biomarkers is done by observing transitions of appropriate groups of predecessor and fragment ion.

The high specificity and understanding are the advantage of targeted proteomics, that enables the collection of proteins in intrinsic samples at concentrations below the ng per mL range. There is some research that is done additionally, in which species-specific peptides for beef, pork, chicken and turkey that are utilized to find out less than 0.5 percent w/w chicken poisoning of pork meat, even after cooking.

There is a classification of parvalbumin as thermostable proteins, and it recognizes the variety of fish also in processed and pre-cooked products. The diverse real commercial samples are used to test the process.

It contains the fastest process for the peptide-biomarker observation that is less than two hours, and its implementation to food-quality control gives the government officials with a quick, actual method for food certification and finding in any food material, that guarantees quality and safety to consumers. In disparity to the growing use of proteomics for the verification of food trials packet, quantitative analysis of food samples by proteomics techniques has been very limited in use.

Also, a label-free quantitative approach has been established using distinctive peptides for quantifying the content of genetic variants of casein in bulk goat milk . Peptidomes may be considered as novel sources of information for food traceability .

The Peptidomics is described as the stream of study that studies and do an assessment of the configuration, relations and properties of the complete endogenous peptide pool of a strong biological system. For example, the use of the process of peptidomics studies to find out the birth of cheese and to detect soybean or milk proteins in other business level protein creation.

There is a possibility that utilization of latest new powerful MS fragmentation modes, such as ETD and HCD, can increase the deep analysis of peptidomes to recognize new possible biomarkers for food certification.

There is a government rule that makes it essential to observe the traceability of raw transgenic materials and the labeling of genetically-modified-organism products. The altered gene means that alteration of endogenous proteins or creation of novel proteins.

The profiling of DNA is utilized to find out genetically modified organism, and the existence of a transgene in a food product is determined by the discovery of transgenic proteins. In short, it can be thought that the utilization of a targeted proteomics-based method will be beneficial for the discovery of genetically modified organism in foods.

9.5 GENOMICS IN AGRICULTURE

The scientific research is done by experiments that can succeed or fail, implementation of genomics can increaseit up to the large level as possible, as a result the betterment of crop quality occurs and an increase in the production of crops can be achieved. The process of crop breeding is used to produce hybrids varieties with the most desirable qualities by connecting features to genes and genes signatures.

The recognition of required qualities by utilizing genomic data by a scientist is done, and then they put these characteristics in a different organism to produce the new genetically modified organism. The crops value and amount of agricultural production can be increased by the application of genomics and scientist are involved in the discovery of such methods.

For example, the innovation in agriculture to produce the drought-resistant crops that can sustain the long dry season with minimal water requirement

9.6 APPLYING PROTEOMICS IN MEAT QUALITY RESEARCH

The scientists can create and test better theories by using biomarkers for growth of muscle and characteristics of meat quality, and the good pointers of quality of meat can profit consumers and meat manufacturing sector. There has been very less research in proteomics, and the implementation of it in meat industry presently is in a nascent stage.

The application of proteome technology to meat and muscle research on livestock animals can give important data during proteome studies of digestion, progress and advancement of muscle tissues in human being and of a mouse.

The knowledge of a variety of softness of meat is essential for the betterment of meat quality. Though many of the biochemical factors are well recognized and the number of features of softness is now known, the main challenge is the intricate methodology of process of cutting the bodies, containing breaking down of protein into chains and relations of dissolvable muscle proteins, intracellular pH and ion movement through the very initial stage of postmortem, and its effect on meat consistency.

The methods of manufacturing and processing technologies can hugely profit from the knowledge of molecular mechanisms that can affect the water-retaining ability of pig meat and the linkage between the features of muscle growth, development, and meat quality.

9.6.1 Growth and Development

The muscle fibers are normally categorized as per their metabolic features as also its contractile properties, and in living beings, muscle fiber features are altered after the birth, comparative to age, sex, and class, and in muscle fibers, the skeletal muscle tissues are the part of it. There has been controversy regarding the association of varieties of fibers and meat quality features like juiciness, taste, and softness, and mainly the impact of fiber forms on meat softness is not yet fully recognized.

In 2005, the experts stated that animal models with detailed documented development phenotypes and special softness features, as well as double muscling in milk-producing animals that are reared by humans, and in sheep the availability of quality of pig have greatly enriched the human beings knowledge of relation between growth of muscles and meat quality.

The scientist that are researching the proteome methods study these models by proteome methods, to gain more knowledge about the molecular methods that associate the features of growth and quality of meat. In 2005, the researcher has presented the first proteomic study of a cattle hypertrophy. They have defined the molecular markers related to an 11-basepair loss in the myostatin gene.

The general levels of inactive myostatin protein get active by mutation. The research conducted had displayed thirteen separately expressed proteins, containing contractile and metabolic proteins and it is compared with proteome patterns found from semi-tendon muscles from young Blue bulls. It appears that the isoform patterns of troponin-T are affected by the myostatin mutation and it is most remarkable.

The myostatin generally regulates infiltration of fast-twitch glycolytic muscle fibers, and it backs up the previous research on bovine muscle hypertrophy, that related muscle hypertrophy with greater infiltration of secondary myoblasts in the fetuses inside the mother womb, and comparatively improved ratios of glycolytic muscle fibers in the adult animals.

In the year of 1981, Boccard stated that muscle hypertrophy also possibly affect perimysium organization, calpain or calpastatin tasks and digestion of fats. In 2002, Mc Pherron and Lee had done more proteome studies particularly with the target of revealing the working process, and it can give information about the association between many muscle growth and features of meat quality.

In 2002, Freking has stated that the Callipyge mutation in sheep is the reason behind muscle hypertrophy due to a mutation in a chromosome that causes the two to three times in calpastatin. The most important part of the calpain or calpastatin is in the softness of meat and it happens at slower speed, also the structural change in muscle from callipyge sheep looks same. Proteome studies of callipyge sheep have not been reported so far, but this will certainly yield useful information. There is possibility of emergence of new knowledge about some of the systems that connect muscle growth, postmortem metabolism and softness in meat but till this time no valid evidence is present.

In 2004, Doherty has documented the Proteome studies of muscle growth in chicken. For the assessment of the role of molecular systems in muscle growth, the chickens are best species due to regular reproduction and different parameters for production of meat in hen and in manufacturing of eggs.

Though the characters of meat quality are not mainly associated to studies of the growth of muscles, the observations in proteome study might in future become essential in knowing the rules of control of skeletal muscle growth, and in studies in future can be associated with important structures of meat quality.

9.6.2 Postmortem Metabolism

The researchers have time and again attracted the attention towards the well-established fact that the biochemical reactions that take place at the time of the initial postmortem period greatly affect the quality of meat. The postmortem metabolism is affected by variations associated with physio-chemical features, histochemical properties temperature, genotypes, and a large number of other factors, though the association to the meat quality has not yet fully known.

The building of models of animals for portraying proteome changes through postmortem storage is a struggling task due to reasons which change the onset of rigor may have a significant effect on softness, containing electrical replication, tension, frightening rate, and nutritional status of the farm animal.

The alteration in proteome at the time of postmortem processes in pork is reported by various experts throughout the years from 2001 to 2004. The pig was killed and in first the first 48 hours of the storage of meat the Postmortem markers are found, it contains structural proteins actin, myosin and troponin and other metabolic enzymes, like myokinase, pyruvate kinase and glycogen phosphorylase, and buildup of particular metabolic enzymes, as well as specific actin and myosin sub-units are examined to associate to meat softness.

The destruction of myofibrillar components is related to the development of softness in meat after postmortem, the responsibility of metabolic enzymes as softness markers has not been in focus yet, but the present proteome information shows that metabolic enzymes might be good biomarkers for the softness of meat.

However, these studies must be regarded as a starting point, and further, the more extensive research study is required to learn about the intrinsic methods that regulate postmortem digestion of muscles and meat, though these studies are only the start.

9.6.3 Calpains Role in Tenderness

In the softness of meat, the calpains have an important role, and that the rate-limiting factor is calpastatin-mediated inhibition of postmortem calpain activity. In other postmortem alterations, it is normal but there is tight meat in callipyge sheep and the less calpain activity after death of sheep. In 2004, there is a research study that showed that transgenic mice that show activities of calpastatin extending the limit generally have some calpain activity after death.

In 2004, another study showed exact calpain assisted disintegration outlines of myofibril proteins.

This type of research was done *in vitro*, by putting together myofibril subfractions with calpain. Specific degradation patterns of the actin, desmin, troponin, and numerous tropomyosin isoforms are found patterns of disintegration. Correlation of these peptide patterns to the softness of meat association with peptide patterns is the further subject of study.

9.6.4 Water Holding Capacity

The essential quality of meat is water-retaining ability, that is closely associated with differences in the postmortem breakdown, but in past ten years the very little development has been observed. There are many single-gene models of porcine drip loss phenotypes that are present, containing the variation of the ryanodine receptor, and variations of the PRKAG3 gene, generally defined as the RN gene.

The Hedegaard in 2004 has documented in a comparative proteome study of the RN-gene effect that the expression data of many enzymes of the glycogen storage paths are differently controlled in a pattern, and the combined data that is obtained from proteome study point out that control of glucose transport is strictly affected in mutant animals.

9.7 FUTURE APPLICATIONS OF PROTEOMICS IN MEAT QUALITY

9.7.1 Markers for Technological Processing

The betterment of processing technologies can be greatly supported by knowing the alterations in protein that occur s in the time of technological processing, and the output is so variable. Numerous studies have shown that after death the glycolytic rates in muscles changes and govern the softness of meat.

The use of electricity in electrical stimulation is broadly utilized to increase postmortem glycolysis; though the system, that is set free by electrical stimulation are not yet known, though slow glycolysis muscles produce tough meat. There is an absence of knowledge on improving the technologies as the current process is generally disordered and harmful to fast glycolysis muscles.

The quality of processed meat through modification of the proteins gets affected by a large number of methods like fermentation, marination, drying, heating, freezing and high-pressure treatment. As with electrical stimulation, the principles and mechanisms behind these processes are often insufficiently understood, hence proteome studies will be of great value in the improvement in meat processing technologies can be achieved by understanding through proteome studies, as with

other methods like electric stimulation is not generally fully known.9.8 The Role of Bacteria in Food Processing and Food Safety

A large number of people get admitted in hospital and some suffer death due to illness caused by consumption of contaminated food. The food poisoning is the main reason behind 325,000 hospitalizations and 5000 deaths in the United States of America, yearly. This is a source of great misfortune but microorganisms and microbial toxins, mainly that are found in food can be utilized as weapons of mass destruction.

In the production process in food technology and biotechnology, the most important factors are cautious monitoring of microbial contamination in the final product and observation of the production process and cleaning and sanitation. The essential analytical problems are the identification, confirmation, and quantification of bacteria and bacterial poison in food.

The *Staphylococcus aureus*, *Campylobacter jejuni,* some *Salmonella* and *Staphylococcus* species, some *Bacillus* strains and *Escherichia coli* O157:H7 strain are the most common bacteria that cause food poisoning. Many complex and effective methodshave been developed to screen the presence ofbacteria and their toxins in food and these methods arelargely based on immunochemical approaches.

The recognition of microbial food impurities and their poisons, for observation of cleaning and sanitation are caused by targeted methods provided by Proteomics and genomics technologies. The alterations of proteomics of poisonous bacteria during food processing and cleaning of devices is observed in research. The latest process of food preservation is the utilization of high hydrostatic pressure technology.

This is a most common knowledge that proteins are the most essential target of high pressure in living organisms and high hydrostatic pressure technology prevents the growth of microorganisms by disabling the key enzymes that are responsible for duplication of DNA and high pressure alsoalters the structure of microbial cell walls and membranes. Though, some bacteria like *Bacillus cereus* can sustain high hydrostatic pressure technology treatment. In the course of high hydrostatic pressure technology treatment, the alteration in the proteome of this model organism has been assessed.

A deadly and probably the most expensive disease that impacts the dairy industry is termed as mastitis, that is caused by infections due to decreased milk production and quality. Mastitis leads to rejection of milk and leads to high rate of cattle mortality. Various researches have been carried out to identify the biomarkers for assessment of the effectiveness of adjunctive therapies used for the reduction of swelling related to mastitis.

Higher expression of some acute-phase proteins such as transthyretin and complement C3 was found in the whey protein samples 18 h after the bacterial infection. And some antimicrobial peptides and more severe phase a-1-acid glycoprotein were also found. The impact of bacterial swelling during mastitis can be indicated by the presence of proteins that can be used as biomarkers for future research. The people involved in the design of cleaning of stainless containers and other surfaces in food processing sites must consider the risk of formation of bacterialbiofilms inside them.

There are some bacteria that can easily live on the surface of stainless-steel containers and other surfaces in the manufacturing facility, even if cleaning and sanitation is done systematically and they are sporogenic bacteria such as *Bacillus cereus*, the Gram- -positive bacterium *Listeria monocytogens* and some pathogenic *E. coli* strains. The contamination caused by these microorganisms can be avoided by a good knowledge of the creation of biofilm and that is behind its destruction. The bacteria that can cause human and animal infection can survive food processing is Staphylococcus species that also build biofilm.

The integration of microorganisms is a type of natural method for their movement, and the high thickness of biofilms provide them good capacity to survive aggressive treatment, but also a suitable biocatalytic possibility. The utilization of stagnant bacterial cells and bacterial biofilms for construction of biosensors for food quality assessment and fermentation process control and use of stagnant yeasts in beer making and winemaking procedures can help reduce the risk of bacterial contamination.

The food processing and storage, interdependence, possible infection and potential food poisoning, their resistance against antimicrobial agents, and the possibility to survive the cleaning and sanitation process can be found out by the precious data obtained from proteomic analysis of biofilm-forming microbial cells.

In the last few years, there is an increased advertisement of food additives that are generally the bacterial species that live inside the human gastrointestinal tract and have been documented in clinical trials. There is another product by the name of probiotics that is being advertised as the cure of all stomach ailments and is defined as *Bifidobacteria* and *Lactobacilli* that are the most popular microorganisms that are added as live bacteria while making food.

There is an increased focus on the survival methods of these bacteria, and on changed protein appearance by bile salt, heat or osmotic attack, which these bacteria are under attack in the human gastrointestinal organs and in the process of production of food.

REFERENCES

1. Bendixen, E. (2005). The use of proteomics in meat science. *Meat Science*, 71(1), pp.138-149.
2. Bongioni, G., Fortin, F., Fusetti, L., Galli, A., Roncada, P. and Greppi, G. (2004). *GENOMICS AND PROTEOMICS APPROACHES FOR CHARACTERISING AND IMPROVING MEAT QUALITY*. [ebook] p.1. Available at: https://pdfs.semanticscholar.org/e63b/2966662a465873d870cc6d70befbc317573a.pdf [Accessed 7 Jun. 2019].
3. Gallardo, J., Ortea, I. and Carrera, M. (2013). *Proteomics and its applications for food authentication and food-technology research*. [online] Research Gate. Available at: https://www.researchgate.net/publication/259099332_Proteomics_and_its_applications_for_food_authentication_and_food-technology_research [Accessed 7 Jun. 2019].
4. Gippe, G., Stecchini, M., Conte, L., Maltini, E., Foti, S. and Cunsolo, V. (2011). *Apoplication of Proteomics in Food Science*. [ebook] p.2. Available at: http://nopr.niscair.res.in/bitstream/123456789/12958/1/JSIR%2070%2811%29%20905-908.pdf [Accessed 7 Jun. 2019].
5. Jagadeesh, D., Kannegundla, U. and Reddy, R. (2017). *Application of Proteomic Tools in Food Quality and Safety*. [ebook] p.2. Available at: http://nexusacademicpublishers.com/uploads/files/AAVS_5_5_213-225.pdf [Accessed 7 Jun. 2019].
6. Opentextbc.ca. (n.d.). *10.3 Genomics and Proteomics – Concepts of Biology-1st Canadian Edition*. [online] Available at: https://opentextbc.ca/biology/chapter/10-3-genomics-and-proteomics/ [Accessed 7 Jun. 2019].
7. Sokac, D., Kovač, S. and Josic, D. (2010). *Application of proteomics in food technology and food biotechnology: Process development, quality control and product safety*. [online] Research Gate. Available at: https://www.researchgate.net/publication/280037787_Application_of_proteomics_in_food_technology_and_food_biotechnology_Process_development_quality_control_and_product_safety [Accessed 7 Jun. 2019].

10

Proteomics: Present and Future in Food Science And Technology

CONTENTS

The present and future of the field of proteomics depends totally on the various factors that affect the progress in the food industry and the development of food technology affects it a lot. It is very important to note that with the advancement of the technology in the field of proteomics, there has been a rise on the number of challenges that were being faced at the functional level. But, with the integration of data analysis with the field of proteomics, it is quite evident to note that there will be certain amount of progress which will help in the overcoming of the challenges.

10.1 INTRODUCTION

Proteomics is the word that is used to explain the observations of the proteins in the experimental sample all over the world. Even though there are various methods might be utilized to perceive and quantify the nature of proteins, the contemporary proteomics now principally depends upon the Mass Spectrometry as the ultimate technique of the selection. This is so because of the extreme level of responsiveness and capability to identify the proteins in very high throughput.

Proteomics is a comparatively modern field of science. The field of proteomics was provided its present title in recent history, specifically in the year of 1994. The term' proteomics' was invented from the aim of the study- Proteins and its superficial analogy with the extensive-ranging and extremely high throughput investigation of the genes and genomics.

The effective completion of the process of the human genetic project has resulted in a marvelous intensification in a better understanding of the molecular foundation of the various diseases. Even though, an all-inclusive understanding of the energetic protein trails included in usual and disease states, and in the response to the medicinal behavior, is needed if the scientist is ready to efficiently diagnose and handle a disease.

Figure 10.1: The field of proteomics has been one of the most basic fields when it comes to the development of the food industry.

Source: http://res.publicdomainfiles.com/pdf_view/194/14021023215756.jpg

The subsequent principal barrier in this direction for this objective is to detect the elements of the human proteome in direction to comprehend the human genome. One specific significance might be to decode the protein changes amongst the well-being and the disease to permit the detection and arranging of the pharmaceutically relevant targets in order of their priority.

Certainly, from the experts' perspective, most of the drug aimed products are mainly proteins and are not made of nucleic acids. Technologies that are accessible to date, for example, microarray might detect a huge number of the distinctly displayed gene structure..

The process of proteome studies has the objectives to recognize the alterations in the expression of the protein structure. Moreover, these proteome studies also focus on the processes of post-translational alterations, protein-protein communications, cellular and sub-cellular dispersal, and temporal outlines of the genetic expression. The resolution of discrepant and practical proteomics is to attain this data that will then spearhead to get an improvised knowledge of the cellular tracks and their inter-relationships in the cellular structures and living organisms.

The influence of proteomics as an instrument for finding the organic paths and the disease procedures is well established in the present times. Undeniably, the process of the proteomics has by now exposed numerous probabilities of finding the modern-day drug goals for changeable diseases. The up-to-date timeline of the process of proteomics is here and now beginning to consider by what means this technology can assist the clinician for high-throughput diagnostic and prognostic submissions.

The usage of the proteomics for the advancement of the procedure and authentication in the progress of food technology and food biotechnology along with the equivalent quality control of the initial materials and the prepared products was rather confined in the initial stages. In the latest stages of the field of proteomics, it has altered swiftly so the proteomics technology is regularly utilized.

There are certain keywords such as the 'industrial process proteomics' and 'industrial proteomics', which are frequently used in the modern world. There are many scientists that have discussed the usage of the genomics and proteomic methodologies for the further advancement, confirmation and optimization of the biological procedures.

In recent times, some experts have also debated the opportunity for the usage of this technology that can be utilized for the authentication of the downstream processing. Apart from the downstream processing, this technology can also be determined for the batch-to-batch differences and superiority controller of therapeutic proteins. Proteomics can also be utilized for authentication and regulator of the manufacturing procedures of the manufactured food goods.

10.2 DIFFERENT APPROACHES IN THE FIELD OF PROTEOMICS

The process of the examination of food products is unceasingly demanding the expansion of extra strong, effective, subtle, and cost-effective logical organizations to assure the security, excellence, and traceability of the food quality in the amenability with the legislature and customers' stresses. The ancient approaches that were utilized at the initial stages of the twentieth century have been grounded on the so-called "wet chemistry".

These ancient approaches have progressed into the present influential methodologies that are utilized in food laboratories in the present times.

This development has resulted in the substantial improvements in the field of analytical correctness, correctness, discovery parameters, and model throughput, by this means intensifying the real-world variety of food applications.

There are many scientists that have mentioned that "the development and organization of the up-to-date worldwide nourishment delivery arrangement totally depend on the process of food analysis. This process of food analysis is basically beyond simple classification. These methodologies are used as the apparatus for modern-day goods growth, superiority controller, monitoring implementation, and problem-solving."

In addition to all this, in the present times, there is an enormous concentration in the health connected possessions of the foods as a consequence of a growing communal apprehension on by what means the health can be improved or enhanced with the assistance of the so-called purposeful food products, functional elements, and nutraceuticals. Thus, there is no doubt on the importance and present-day requirement of diagnostic methods advancements that might be able to confront all these requirements.

Conventionally, the investigative methodologies have been divided on the basis of their functioning principle. For instance, these methodologies can be spectroscopic as in the case of mass spectrometry (MS); nuclear magnetic resonance (NMR); infrared (IR); atomic spectroscopy (AS), biological (polymerase chain reaction (PCR); immunological techniques; biosensors), electrochemical methodologies which also include biosensors.

Moreover, these investigative technologies can be used for separation purposes. Some of these technologies for separation are high-performance liquid chromatography (HPLC); gas chromatography (GC); capillary electrophoresis (CE); supercritical fluid chromatography (SFC). In separation, the methodology for sample preparation, there are investigative technologies like the solid phase extraction (SPE); supercritical fluid extraction (SFE); headspace (HS); flow injection analysis (FIA); purge and trap (PAT); microwave-assisted extraction (MAE); automatic thermal desorption (ATD). In another methodology, there is a type of investigative technology that is named as hyphenated. In this kind of investigative methodology, the separation and spectroscopy technique are used together on the food samples.

Figure 10.2: There are different approaches in the field of proteomics that are being used in the present day.

Source: https://live.staticflickr.com/8002/6993514266_036054c903_b.jpg

Every possible methodology delivers a particular set of data on the specimen or sample that has been put under observation. These samples have been put under observation that is based on a physical-chemical reaction. All these specimens exhibit certain properties during the progression of the chemical and physical reactions. And also have their own positive and negative effects when chemical and physical reactions are applied to these food specimens in the process of food analysis.

There are many reports and studies that have described many logical methodologies that are being frequently used in the process of food analysis. These reports have presented further knowledge on the difficulty of the various methodologies that are being currently used in the process of food analysis. This further knowledge on the various methodologies can be attained by progressing a little more on any one of the conventional methodologies that are being currently used.

There are many different cases of the varied immunological methodologies. These various immunological methodologies comprise of several methodologies. Some of the major immunological methodologies are enzyme immunoassay (EIA), enzyme-linked immunosorbent assay (ELISA), immunodotting, radioimmunoassay (RIA), solid-phase RIA, liquid-phase RIA, immunoradiometric assay (IRMA), fluorescence: fluorescence immunoassay, enzyme-linked fluorescent immunoassay, fluorescence polarization immunoassay (FPIA), time-resolved fluorescence immunoassay (TRFI), and chemiluminescence immunoassay (CIA). Along with these methodologies, there are various other methodologies. The total number of the methodologies that are included in the immunological methodology is near about twenty-seven s. These techniques are used in different combinations with each other to successfully form the various immunological methodologies.

There are many studies and researches that have laid their focus on the modern-day advancements and usages of the innovative and active analytical methodologies that can be used to identify the composites that are of importance for the field of food science and technology.

In the modern world, there are many scientists who have been working on various projects that are totally focused on new-age analytical mechanisms. These scientists have focused primarily on the expansion of the innovative methodologies and their area of usage in the field of food science and technology.

In the upcoming time, these researches are directed towards the study in the field of Foodomics. Along with the development of the new-age fields of food science and technology, the experts and scientists have directed their researches and present-day studies on the aspect of the quality controller and security, nourishing cost, processing impacts, storage, bioactivity, and different aspects of food science and technology.

The observation and study of a huge diversity of the molecules, that related to food, with the various biochemical attributes has been considered by various scientists. Some of the molecules that are in consideration for researches, due to their connection with the food technology, are amino acids, peptides, proteins, phenolic compounds, carbohydrates, DNA fragments, vitamins, toxins, pesticides, additives, and chiral compounds. Apart from this, a worldwide scenario on the progression of the major analytical methodologies that are being used in the field of food science and analysis is also on the to-do list of the various experts and scientists.

There is a huge number of publications and studies that have been conducted in the previous two decades. According to a survey, the publications that were published in the first decade of the twenty-first century has been compared with the publications and the studies that were published in the last decade of the twentieth century on the topic of food analysis and proteomics.

10.3 HISTORICAL ADVANCEMENTS IN THE FIELD OF PROTEOMICS

10.3.1 History: From farm to factory

Amongst the sixteenth century and in the initial years of the eighteenth century, there was a very little advancement in the methodologies of the farming that were used at that time. In the sixteenth century, there was the advent of the usage of innovative and new tools in the field of farming. Some of the new tools that were used at that time were the horse-driven hoe and field drill.

This was the first horse-driven hoe and drill that was discovered by the scientist named as Jethro Tull. Apart from farming, the rearing of the livestock progressed. In the animal cultivation and livestock rearing, some of the of new breeds and species of the cattle were introduced. The new species were very productive and gave a greater quantity of milk yields.

10.3.2 Before the Industrial Revolution

From the sixteenth century to some initial years of the nineteenth century, the commodity of food was majorly cultivated by the manual methods and then it was either cooked or eaten raw in many areas of the world. With the advent of food technology, some of the elementary food articles, for example, bread, were being cooked in the bakeries and were made available for the people to consume.

As a matter of fact, the cost and constituent materials of bread were controlled by the royal mandate, even in the year 1276 AD. At that time, the manufacturing phases of bread were at household level. The manufacturing units and production plants of bread were simply based on the house kitchens. In these types of kitchens, the bread was baked in ovens and on open fire structures.

10.3.3 The Industrial Revolution

With the dawn of the age of the Industrial Revolution, many different technologies were clubbed together. The clubbing of these technologies has opened the way for the innovative potentials for the examination. The process of Automation occurred in this age. This automation has assisted in enhancing the quantity of the manufacturing of goods, especially food products.

During the Industrial Revolution, there was an increase in the number of factories that were being established. The establishment of more and more factories has led to an increase in the process of mass production of the products. With the establishment of the factories for mass production, the establishments of the townships and cities also started expanding. Eventually, this explained that there were a greater number of people that were to be fed and required a proportionally increase in the number of factories that were being established.

10.3.4 The Recent Future

Progressively in the past ten years, the field of proteomics has proved to be an exclusively utilized methodology in the field of cell biology and food technology. The increase in the role of the proteomics in the field of food science and technology is not only due to the certain technological advancements in the analytical mechanisms that were being used in the field of food science and technology. This increase in the usage of the proteomics is because of the innovations ofdifferent methods which are used to study the mechanisms of protein and peptide related to food.

This extensive usage of the proteomics in the field of food technology has given rise to the expansion of usages and possibilities of the utilization for proteomics. The further usage of the proteomics involves the examination of the cellular structures and tools to the process of imaging of the protein circulations in the zones of cells and tissues.

The increasing measure and the possibility of the contemporary proteomics developments have brought to the explicit computing challenges that are included in handling and extraction. This results in the formation of extremely huge and complicated sets of information.

10.3.5 Contemporary Proteomics

The field of Contemporary proteomics is hence characteristically multidisciplinary. This requires the proficiency of the software technicians and mathematicians along with the experts in the fields of cell biology, protein chemistry, and mass spectrometry. There are many studies that have described some of the sophisticated, individual, and innovative usages of the mass spectrometry-based proteomics to the areas of medical research, diagnostics, quality control, and drug discovery.

10.4 PROTEOMICS AND DIFFERENT FOODSTUFFS

10.4.1 Meat and Meat Foodstuffs

As described by the customers, tenderness, succulence, essence and color are the most significant quality aspects of meat. Particularly for pork and poultry, the water-holding volume is also considered as the most vital quality trait that is thoroughly connected to postmortem variations in meat proteome.

These variations have currently been examined by the usage of proteomic approaches. The content and structure of the intramuscular connective muscle are another, often abandoned feature that expressively impacts the quality of the meat, particularly the tenderness of the prepared meat.

The content and comparative concentration of diverse kinds of collagen differ and are reliant on the variety of meat product. The process of proteolysis is accountable for deprivation of collagen and other constituents of intramuscular connective muscle, for instance, proteoglycans, during the postmortem stowage of the meat product.

This process is also partly accountable for the improved tenderness of meat after the process of conditioning. Turnover of connective tissue, particularly the turnover of dissimilar categories of collagen appears to be measured by matrix proteinases and their definite inhibitors.

Though, the part of these proteins in regulating the quality of the meat still requires to be explored in a better way. In the end, the connective tissue that is comprising of the dissimilar categories of collagen, proteoglycans and other matrix proteins are accountable for the so-called 'background' characteristics of the meat.

Therefore, the further recognition and quantitation of collagens and collagen proteolysis, also by the usage of proteomic approaches, is essential. The intramuscular connective tissue is just one of the features that have an impact on the tenderness and succulence of the meat foodstuffs.

Figure 10.3: The meat industry has grown a lot with the help of Proteomics.

Source: https://c.pxhere.com/photos/a9/ca/cow_cattle_highland_scotland_show_livestock_agriculture_beef-1254724.jpg!d

These characters are further impacted by both heredities as well as the ecological aspects, and pre- and postmortem actions. The information that is attained by discovering and understanding the differences in the proteome in relation to certain superiority and/or processing parameters can be effectively used in order to enhance the making of the meat products.

According to Hollung et al. and Bendixen, in meat science and technology proteomics can be utilized for:

- Proteome mapping and the identification of the meat products
- Determination of proteome variation because of the hereditary differences
- Determination of variations because of the pre-slaughter situations
- Determination of postmortem variations
- Research of variations that are linked with the quality traits of meat

- Peptidomics – research and discovery of the variations in peptide configuration at the time of the storage and processing of the meat products;
- Quality evaluation of dry-cured and fermented meat foodstuffs. There exist some food security concerns, particularly toxicological and microbiological security of meat and meat foodstuffs, but also some other points concerning the use of proteomics in the recognition of the currently known safety dangers, such as avian and swine flu.

10.4.2 Milk Industry

Milk proteins offer the most important nutrition for the development and it signifies an extensive range of functional food constituents. Majority of proteins are carefully consumed in the gut in order to offer the important vital amino acids, though, others are just partly or slightly fragmented down and therefore able to exert higher-level functionality associated to the structures of their digestive foodstuffs.

Describing the milk proteome and how it fluctuates at the course of lactation are the important stages in the direction of an enhanced understanding of milk biology and function and has the capability to offer the innovative understandings in the fields of dairy and other food sciences.

This improved understanding may also help in supervising the manufacturers in the development of foods with better-quality protein and function. The current advancement in proteomics has significantly enlarged the number of proteins that are recognized in milk.

The current discoveries have been highlighted regarding both human as well as bovine milk proteomes that are innovative or whose significance is not yet entirely understood.

Also, there is a description of the use of proteomics in describing the heat-induced protein alterations during industrial procedures that frequently decrease the nutritional value and purpose of the milk proteins.

Post-genomic apparatuses and technologies have intensely altered the investigational methods by which composite biological systems can be categorized. The method of proteome examination using two-dimensional polyacrylamide gel electrophoresis (2D-PAGE) and mass spectrometry has the ability to control the worldwide variations that happen in the protein complement of tissues and subcellular sections.

An innovative challenge for proteomics has currently been known in the variation of food proteomes. There were many reports which pay attention to the current information and the upcoming potential utilization of proteomics in food science and technology.

Proteomics not just has the innovative inferences for organic and forensic usages, it is pushing the restrictions of calculation, record generation and curation, and information. The usage of high throughput methods in proteomics, has caused in the researches that produce large volumes of data. For instance, presently one sample, operate on a mass spectrometer for ~100 minutes, can make the raw data records up to 2.5GB.

This amount of data generation rapidly builds up to a very great scale and given the on-going enhancements in instrumentation this trend is just probable to rise in the upcoming times. At the present times, there exist many funding bodies that are supporting proteomics investigation have rules that require the storage of this information for a least of ten years.

Simple storage of specific information records is not just the challenge, though, particularly as the price of large-scale storing devices continues to drop down. The actual challenge is to create very large volumes of data accessible for the continuous examination without any difficulty. The activities supporting this concept comprise of the PRIDE database. This database is communal with MS data to serve as a statistical tool for significance.

10.5 PRESENT AND FUTURE OBJECTIVES OF PROTEOMICS

The main objective of proteomics is to classify all proteins, their organic action, post-translational alterations and connections within a cell, and to recognize (quantify?) the variations in «proteome» in response to the changed biological situations.

A typical proteomics flow of work comprises of the protein abstraction, separation, protein or peptide recognition and study of the information. Mass spectrometry is considered as the most common technique. This is used to find out the presence of proteins or peptides in proteomics.

This technique has several numbers of applications, which involves the investigation in meat science, but it is restricted by the vast biochemical heterogeneity of the proteins and an incapability to sense the correctly low-abundance proteins that are present. The objective of the current evaluation is to precise the recent information and recognize the upcoming potential application of proteomics in meat science and technology.

Proteomic methods are categorized into six collections: expression proteomics, protein–protein connections, functional proteomics, structural proteomics, proteome mining and post-translational alterations.

Proteomics is considered as an innovative promising method. This is used to recognize the protein in the food matrix and to investigate the protein–protein connections in both uncooked as well as processed foods, as well as the connections between proteins and other components of the foodstuff.

It offers a sensitive data on the variations in protein structure occurring at definite amino acid residues during the actions of processing and helps to forecast the quality and constancy of food product.

In proteomics, an examination of proteins and peptides is the most important component and the most challenging process is their separation from the complication and dynamic concentration range in the food products.

The most important application of proteomic method is the research on the variations in the quality of the food products, recognizing the bioactive composites, recognizing the components in functional foodstuffs and nutraceuticals (Galvani et al., 2001), classification of rice proteins (Koller et al., 2002), molecular indicator for physiology and pathobiology (Bouley et al., 2005).

10.6 VARIATIONS OF PROTEOMICS RESEARCH

It is well recognized that several different protein alterations such as phosphorylation, oxidation, and degradation can impact the quality of the meat products and it is thus vital to execute and progress the proteomics approaches in meat science to recognize the mechanism behind the impact of these protein alterations in the quality of meat and meat products.

It can be assumed that that high-throughput modification-specific proteomics will be employed in order to examine methodically the qualitative as well as the quantitative alterations of these protein alterations in meat regarding the growth, hereditary background, processing, and storage that will significantly effect the reorganization of the mechanisms which emphasize the meat quality alteration.

The proteomics research that is made on the post-mortem variations in meat evidently demonstrate great potential in the study of meat. Proteomics has particularly shown to be an influential device to examine the post-mortem protein deprivation in meat because it offers valued data regarding the composite mechanisms which are behind the post-mortem proteolysis.

Though, it is significant to emphasize that the proteome of a muscle cell or a meat product is enormously composite in nature and there exist not a single proteomic technique that can profile a whole proteome.

In addition to it, all approaches in proteomics are very complex and along with it, these techniques are time-consuming and just a restricted number of examples can be examined in each research. Henceforth, it is often essential to do numerous numbers of different proteomics researches in order to disclose a definite mechanism.

10.7 MASS SPECTROMETRY

The data sets of mass spectrometry unaided cannot deliver any resolution to the main queries in the field of applications of the proteomics that are in the use. For gaining access to the data sets of Mass spectrometry, the data sets must be first understood and examined. The procedure of changing the Mass Spectrometry statistics to data and then analyzing and converting it to applicable knowledge requires sophisticated methods of data processing and analytics.

These approaches should essentially consider the various multi-faceted, compound nature of the international study of the structures of proteomes. These studies should include the parameters such as what quantity of a protein is present in a cell. This quantity should be measured in the aspects of concentration and ampleness.

Figure 10.4: The process of Mass Spectrometry has been one of the most frequently used methods in the field of Proteomics.

Source: https://live.staticflickr.com/8523/8469357737_eeeffded1b_b.jpg

Apart from quantity, the next parameter is the location of the protein in the cell. It is very important to note that where the protein is placed in a cell or what is the spatial localization of the cell. After the location, the other important factor is the partners of interactions.

It is very important for the experts to know that with whom does the protein interacts or 'with whom does the protein 'talks' to. From this, the experts can also judge what kind of alterations are possible in the protein on the chemical level and what can the protein gain or lose after the chemical reactions.

Henceforth, once after storing these sets of data, the convectional linked databases can be considered, however, more recently multi-dimensional databases have been shown to play a role in providing a structure for proteomics data that allows for fast dimensional analytics to take place.

10.8 PRESENT-DAY PROGRESS OF PROTEOMICS

There was a group of scientists that displayed an evaluation of the set of data and certain storage platforms for the experiments of immuno-precipitation. There were some opinions to an evolving movement in the field of proteomics. For example, the usage of the huge sets of databases which were linked with high-class and developed assessment software.

The Boulon paper defines the primary application of commercial intellect policies in the field of proteomics to help in the process of analysis of the immuno-precipitation experimentations. An immuno-precipitation experimentation is the besieged separation of a protein that is in focus, with the supposition that organically networking proteins will get purified with the other proteins as well.

This is the scenario, but most frequently various random strands of proteins also get purified with the other proteins in just about all the experimentations. Putting the statistics that is connected to the rate of recurrence of detection of proteins in use in the process of multiple immuno-precipitation experimentations aids to expect which kind of protein recognitions might not be true.

TA certain group of scientists explains a policy of computationally joining experimentations to produce a rate of recurrence of the detection for the proteins in several immuno-precipitation experimentations.

It would be very helpful touse this source as it aids in assessing if a protein recognition in innovative experimentation might be a non-specific protein or a genuine interaction partner. This application of business intelligence policies delivers evidence of the idea for the joining of the number of different experimentations in an organized set of data store. This is done in order to help the statistical assessment of the upcoming experimentations.

Moreover, the use of certain computational evaluation methodologies in this direction allows many experts to join large volumes of data. This helps them in better addressing the complication of proteomics, and the usage of the ancient information that has been discovered in the evaluation of the modern-day experimentations.

The usage of the proteomics that has been discussed explains how the proteomics plays the role of indicator in the modern-day experimentation. The usage of proteomics explains how capable or potent is the proteomics if it is joined with the area of data analysis. The fast-moving proteomics, when linked with the field of data analytics, might become capable of assisting the process of improvisation of the usage of the high throughput multi-dimensional sciences like the fields of genomics and proteomics.

According to many experts and scientists, if an attempt is made to investigate the future of proteomics, then the upgraded level of data handling, analysis and examination methodologies will be very common amongst the processes that are going to be used in the future. One of the main reasons for this transition is the non-feasible physical experimentations, which are no longer feasible in the field of proteomics and might get vanished from the scene of food technology and experimentation.

Thus, to maintain the functionality of the field of proteomics and the food technology, it is quite predictable that this level of data management and examination must turn out to be very commonplace in the food industry and proteomics. Additionally, accumulating the sets of data methodically can assist to create a reference point of the historical knowledge that can be retrieved and used to discover certain designs that are visible in various experimentations over the period of time.

10.9 FUTURE PROSPECTS AND CHALLENGES OF PROTEOMICS

The field of Proteomics is analogous to the easier form of evaluation. Some of the hard problem solving includes arranging a limited quantity of fragments together to disclose the multiple responses. This is done because usually there is a single result or outcome of the problems of the proteomics.

The field of Proteomics and several another great level high throughput methods are extra multifaceted in their form of analysis. This is so because irrespective of the number of variables that are used in the process of analysis, the results of the entire analysis will amount to the hundreds to thousands.

There is a need for the shrewd calculation. This clever or shrewd computation usually be made up of the database handling and analytics answers for managing this level of complexity. This level of computation is very effective and will be needed more often in the future to extract the most out of the data.

Proteomics as an evolving, and by nature high-tech, multidisciplinary science is having an extensive impact on the human race. Appreciations to many of the complementary progressions in practice, cutting edge physical chemistry and engineering along with the power of the recent computing, proteomics also provides massive future potential for responding to the most important queries in the field of biology.

In addition to it, it also offers innovative clinical signs of progress, which further give rise to enhanced analyses and the treatment of human disease without any harm. The use of proteomics in these innovative areas, for instance forensics, medication discovery, evolutionary investigation, and food security is a signal of the increasing significance of this type of investigation.

The privileges of customers and unaffected food processors in the aspect of food contamination and deceitful or misleading methods in the field of food processing have already been pre-decided by the various organizations of the world. There are many bodies that have already formed the guidelines concerning the security and tractability of the food products.

The system and the guidelines for the food safety were developed in order to:

- Inspire the agricultural production of different food crops
- Guard the product names and signs from being misused and imitated
- Aid the customers by providing them the set of data that is related to the particular attribute of the food product

The probable use of the field of foodomics in the progressive path of the food production, evaluating the security, novelty and class of the food product. As the cohesion of the foodomics with the food technology was strengthened, it visibly appeared that the incorporation of all foodomics topics will signify a more appropriate tool than putting a single field of foodomics in use. For example, if the field of proteomics or genomics alone is used, then it would not be that beneficial as when different fields of foodomics are combined.

The outcome was an exhaustive association amongst the specialists on dissimilar areas of food science and nourishment and omics specialists. The usage of foodomics allowed, or at the slightest effect, enabled the responses to the queries that are critical for food security, eminence, tractability and its defense against the practice of food contamination.

These points are:

- topographical source
- management
- eminence and innovativeness of the raw material used
- usage of inherently altered food and/or transgenic bacteriological practices
- usage of antibiotics, fungicides and herbicides throughout the process of animal rearing and food crop agriculture

- bacterial adulteration during the manufacturing process, management and storage.

The usage of the proteomics for course progress and authentication in the field of food technology and food biotechnology, as well as consistent quality regulation of the raw ingredients and the processed or ultimate goods, was, in the starting, somewhat restricted. There were only some of the scientists all over the world that had focused on this aspect of the issues. There were many seminars and presentations that were carried in consideration to the sections of Biotechnology perspectives' and 'Proteomics in biotechnology' at the HUPO World Congress near about half a decade ago.

Out of these seminars and presentations, only some of them are dedicatedly trading with the usages of the proteomics. It has been noticed that in some of the previous years, the field of proteomics has evolved at a very fast speed. There are many different terms that were being used in proteomics technology but have changed in the routine usage.

For example, there are some terms such as the 'industrial process proteomics' and 'industrial proteomics' are some more words that are being regularly utilized. Some scientists have also debated the usage of the genomics and proteomics methods for expansion, authentication and optimization of the certain bioprocesses.

Lately, there are some more scientists that have also explained the opportunity for the usage of this technology for the authentication of downstream processing, the resolve of batch-to-batch dissimilarities and eminence regulation of beneficial proteins. The field of Proteomics can also be utilized for justification and regulation of the various manufacturing procedures of the food products. In a revolutionary assessment, some scientists have explained how the field of proteomics can be used as an instrument for the study of the seafood excellence and recognition of the conceivable microbial adulteration.

The subsequent premature usage of proteomics in food technology and for quality control was the evidence of usage of the anabolic steroids in the products of meat and milk. By the usage of the 2D electrophoresis and matrix-supported laser desorption/ionization-time of flight (MALDI-TOF) mass spectrometry, some scientists and experts recognized numerous applicants for quality indicators for the postmortem alteration of the muscle in pork meat within the time period of storage. The main challenge in the usage of proteomics in the food industry is totally grounded on the dispensation of the raw material that has to be used in manufacturing plants.

The main issue is that the entire genome arrangement of numerous species of various families of plants is still not known and must be discovered. This condition is nowadays rapidly improving, and the genome of plants such as rice that are significant for the human and animal nutrition is here and now either sequence, or their sequencing is the main concern for the ongoing projects.

The analysis of the alfalfa Medicago sativa protein pattern was carried out. During the manufacturing dispensation, some experts had explained the usage of the proteomics that is available for procedure expansion and quality control. The genetic structure of this plant was not yet formed into a sequence. Apart from this, the set of data that was accessible from the linked genomes had to be used.

REFERENCES

1. Have, S., Ahmad, Y. and Lamond, A. (2013). Proteomics - Current Novelties and Future Directions. *Journal of Analytical & Bioanalytical Techniques*, [online] 04(03). Available at: https://www.omicsonline.org/proteomics-current-novelties-and-future-directions-2155-9872.S3-001.php?aid=1955 [Accessed 10 Jun. 2019].
2. Josic, D. and Giacometti, J. (2013). *Foodomics-Use of Integrated Omics in Nutrition, Food Technology and Biotechnology*. [ebook] Rijeka: Journal of J Data Mining in Genomics & Proteomics. Available at: https://www.longdom.org/open-access/foodomics-use-of-integrated-omics-in-nutrition,-food-technology-and-biotechnology-2153-0602.1000e106.pdf [Accessed 10 Jun. 2019].
3. Zargar, S., Mahajan, R., Nazir, M., Nagar, P., Kim, S., Rai, V., Masi, A., Ahmad, S., Shah, R., Ganai, N., Agrawal, G. and Rakwal, R. (2017). Common bean proteomics: Present status and future strategies. *Journal of Proteomics*, [online] 169, pp.239-248. Available at: https://europepmc.org/abstract/med/28347863 [Accessed 10 Jun. 2019].
4. Cifuentes, A. (2012). Food Analysis: Present, Future, and Foodomics. *ISRN Analytical Chemistry*, [online] 2012, pp.1-16. Available at: https://www.hindawi.com/journals/isrn/2012/801607/ [Accessed 10 Jun. 2019].
5. Carbonaro, M. (2004). Proteomics: present and future in food quality evaluation. *Trends in Food Science & Technology*, [online] 15(3-4), pp.209-216. Available at: https://eurekamag.com/research/004/287/004287257.php [Accessed 10 Jun. 2019].
6. Gaso-Soka, D., Josic, D. and Kova, S. (2019). *Use of Proteomic Methodology in Optimization of Processing and Quality Control of Food of Animal Origin*. [ebook] Rijeka: Food Technol. Biotechnol. Available at: https://pdfs.semanticscholar.org/23e5/962e2db8b6ad3349ebe1f2598ded300b6aa2.pdf [Accessed 10 Jun. 2019].
7. Han, J. and Wang, Y. (2007). Trends in Food Science & Technology. *Elsevier Ltd.*, [online] 19(1), pp.26-30. Available at: https://www.sciencedirect.com/science/article/abs/pii/S0924224407002294 [Accessed 10 Jun. 2019].
8. Han, J. and Wang, Y. (2019). *Proteomics: present and future in food science and technology*. [online] AGRIS: International Information System for the Agricultural Science and Technology. Available at: http://

agris.fao.org/agris-search/search.do?recordID=US201300841547 [Accessed 10 Jun. 2019].

9. Have, S. (2013). Proteomics - Current Novelties and Future Directions. *Journal of Analytical & Bioanalytical Techniques*, [online] 04(03). Available at: https://www.omicsonline.org/proteomics-current-novelties-and-future-directions-2155-9872.S3-001.php?aid=1955 [Accessed 10 Jun. 2019]
10. Jagadeesh, D., Kannegundla, U. and Reddy, R. (2017). *Application of Proteomic Tools in Food Quality and Safety*. [ebook] Chennai: nexus academic publishers. Available at: http://nexusacademicpublishers.com/uploads/files/AAVS_5_5_213-225.pdf [Accessed 10 Jun. 2019].
11. HAN, J. and WANG, Y. (2008). Proteomics: present and future in food science and technology. *Trends in Food Science & Technology*, [online] 19(1), pp.26-30. Available at: https://www.researchgate.net/publication/222204081_Proteomics_present_and_future_in_food_science_and_technology [Accessed 10 Jun. 2019].
12. Kova, S., Josic, D. and Gaso-Soka, D. (2010). *Application of Proteomics in Food Technology and Food Biotechnology: Process Development, Quality Control and Product Safety*. [ebook] academia. Available at: https://www.academia.edu/23335837/Application_of_Proteomics_in_Food_Technology_and_Food_Biotechnology_Process_Development_Quality_Control_and_Product_Safety [Accessed 10 Jun. 2019].
13. Slideplayer.com. (2019). *Food Science & Technology Past Present & Future - ppt download*. [online] Available at: https://slideplayer.com/slide/7425758/ [Accessed 10 Jun. 2019].
14. Toldrá, F. and Nollet, L. (2013). *Proteomics in Foods Principles and Applications*. Springer US.
15. Toldrá, Fidel, Nollet and Leo, M. (2013). *Proteomics in Foods*. New York: Springer US.
16. Verrills, N. (2007). *Clinical Proteomics: Present and Future Prospects*. [online] PubMed Central (PMC). Available at: https://www.ncbi.nlm.nih.gov/pmc/articles/PMC1579414/ [Accessed 10 Jun. 2019].
17. Verrills, N. (2019). *Clinical Proteomics: Present and Future Prospects*. [online] PubMed Central (PMC). Available at: https://www.ncbi.nlm.nih.gov/pmc/articles/PMC1579414/ [Accessed 10 Jun. 2019].

18. Zamaratskaia, G. and Li, S. (2017). PROTEOMICS IN MEAT SCIENCE — CURRENT STATUS AND FUTURE PERSPECTIVE. *Theory and practice of meat processing*, [online] 2(1), pp.18-26. Available at: https://www.ingentaconnect.com/content/doaj/2414438x/2017/00000002/00000001/art00002 [Accessed 10 Jun. 2019].

INDEX

D

E

F

G

P

Q

R

S

T

U

V

W